205

Topics in Current Chemistry

W0049938

Springer-Verlag Berlin Heidelberg GmbH

Organosulfur Chemistry II

Volume Editor: Philip C. B. Page

With contributions by
N. Furukawa, R. S. Glass, J. Nakayama,
S. Sato, Y. Sugihara

 Springer

This series presents critical reviews of the present position and future trends in modern chemical research. It is addressed to all research and industrial chemists who wish to keep abreast of advances in the topics covered.

As a rule, contributions are specially commissioned. The editors and publishers will, however, always be pleased to receive suggestions and supplementary information. Papers are accepted for "Topics in Current Chemistry" in English.

In references Topics in Current Chemistry is abbreviated Top. Curr. Chem. and is cited as a journal.

Springer WWW home page: http://www.springer.de
Visit the TCC home page at http://link.springer.de/

ISSN 0340-1022
ISBN 978-3-662-14712-2   ISBN 978-3-540-48986-3 (eBook)
DOI 10.1007/978-3-540-48986-3

Library of Congress Catalog Card Number 74-644622

Cover design: Friedhelm Steinen-Broo, Barcelona; MEDIO, Berlin
Typesetting: Fotosatz-Service Köhler GmbH, 97084 Würzburg

SPIN: 10694584   02/3020 – 5 4 3 2 1 0 – Printed on acid-free paper

Topics in Current Chemistry
Now Also Available Electronically

For all customers with a standing order for Topics in Current Chemistry we offer the electronic form via LINK free of charge. Please contact your librarian who can receive a password for free access to the full articles by registration at:

http://link.springer.de/series/tcc/reg_form.htm

If you do not have a standing order you can nevertheless browse through the table of contents of the volumes and the abstracts of each article at:

http://link.springer.de/series/tcc

There you will also find information about the

- Editorial Board
- Aims and Scope
- Instructions for Authors

Preface

Organosulfur Chemistry has enjoyed a renaissance of interest over the last several years, fuelled by its impact in the areas of heterocyclic and radical chemistry, and particularly stereocontrolled processes including asymmetric synthesis. One result of this resurgence of interest in the field is a rapidly escalating number of related publications. These volumes are intended to provide coverage of some of the highlights of contemporary organosulfur chemistry chosen from the entire range of current activity.

The first volume begins with a comprehensive review by Prof. José Luis Garcia Ruano and Dr. Belén Cid de la Plata of asymmetric cycloaddition mediated by sulfoxides, including dipolar and other processes in addition to Diels-Alder chemistry. It is followed by a discussion of the synthetic uses of thiocarbonyl compounds by Prof. Patrick Metzner.

Volume 2 begins with a thorough survey of sulfur radical cations, covering their synthesis, structure, stability, and reactivity, by Prof. Richard Glass. Prof. Naomichi Furukawa and Prof. Soichi Sato describe recent developments in the area of hypervalent organosulfur compounds, and the volume is completed by a discussion of the chemistry of thiophene 1,1-dioxides by Prof. Juzo Nakayama and Prof. Yoshiaki Sugihara.

Leicestershire, July 1999   Philip C. B. Page

Contents

Contents of Volume 204

Organosulfur Chemistry I

Volume Editor: Philip C. B. Page
ISBN 3-540-65787-8

Sulfur Radical Cations

Richard S. Glass

Department of Chemistry, The University of Arizona, Tucson, AZ 85721, USA
E-mail: rglass@u.arizona.edu

Sulfur radical cations are novel reaction intermediates that have attracted considerable attention recently. This in part is due to the renewed interest in the chemistry of radical cations in general, as well as four other factors. The first is the basis for understanding the structure and reactions of radical cations – can their inherent nature be best depicted by analogy with radicals, with cations, or is a new algorithm required? As is exemplified in this review, sulfur radical cations show novel behavior. Second, some reactions of sulfur radical cations are attracting interest for their application in organic synthesis. Third, biological electron-transfer may be mediated by sulfur and, indeed, sulfur radical cations may be intermediates in biological redox processes. Finally, materials based on tetrathiafulvalene, polythiophene, and related sulfur compounds have high electrical conductivity which may be understood in terms of p-delocalized sulfur radical cations or the related dications. Consequently, a comprehensive and critical review of this field appeared timely.

Keywords: Reactive intermediates, sulfur radical cation structure preparation and spectroscopy, π-delocalized and localized sulfur radical cations, 2c3e bonds, di- and polysulfide radical cations, biological oxidation of thioethers, distonic ions

Topics in Current Chemistry, Vol. 205
© Springer-Verlag Berlin Heidelberg 1999

1
π-Delocalized Sulfur Radical Cations

1.1
Thianthrene Radical Cation

Thianthrene radical ion (1+), thianthrene radical cation **1**, has played an important role in radical ion chemistry because of the ease of oxidation of thianthrene (TH) and the stability of its solid salts.

1

Although this species had long been known [1] due to its formation on dissolving thianthrene in concentrated sulfuric acid, its structure was not correctly elucidated until revealing analysis of its EPR spectrum was accomplished by Shine and Piette [2], Kinoshita and Akamatu [3], and Lucken [4]. An engaging personal recounting of this important development has been published [5]. The EPR spectrum of thianthrene radical cation in sulfuric acid shows five lines indicating hyperfine splitting by four equivalent protons and an average g-value of 2.0081 consistent with substantial spin density on sulfur resulting in a larger g-value than the free electron value due to spin orbit coupling. EPR spectroscopic studies on the radical cations obtained from substituted thianthrenes [6] clearly showed that the four proton splitting in the parent was due to H2, H3, H7, and H8. Generation of thianthrene radical cation using aluminum chloride in nitromethane provided a better resolved spectrum than that in sulfuric acid, in which the smaller hyperfine splitting due to H1, H4, H6, and H9 ($a_H = 0.135$ G) versus that of H2, H3, H7, and H8 ($a_H = 1.28$ G) could be measured as well as the hyperfine splitting due to natural isotopic abundance S-33 [7]. The g-value and S-33 hyperfine splitting provided proof of spin density on the sulfur atoms. The distribution of the spin over both sulfur atoms and C2, C3, C7, and C8, principally, suggests that thianthrene radical cation is the resonance hybrid of contributing structures **2a–d** as well as those shown in **1**. Such delocalization implies that this species is planar.

| **2a** | **2b** | **2c** | **2d** |

This is in contrast to the nonplanar geometry of thianthrene. The solid state structures of thianthrene and a number of its derivatives have been determined by X-ray crystallographic techniques. In general these compounds adopt a non-planar C_{2v} conformation with the central ring in a nonplanar boat conformation with a dihedral angle Θ, as shown in **3**, of 127–128° for thianthrene [8], 127–134° for 2,3,7,8-tetramethoxythianthrene [9] and an overall range of 125–143° for other derivatives [10–18].

3

This is in agreement with recent molecular orbital [19] and PM3 calculations [20, 21] in which it was found that thianthrene has a double minimum potential for the folded conformation but the barrier between them is very low (less than 1 kJ/mol [18], 0.67 kcal/mol [19], 1 kJ/mol [20]). Thus the interconversion via the planar form is not strongly disfavored despite the formal antiaromaticity of the planar conformer. Indeed it has been suggested that the major contributor to the higher energy of the planar form of thianthrene is bond angle distortions in the central ring [18]. Thianthrene derivatives **4–6** adopt planar conformations in the solid state as shown by X-ray crystal structure studies [21–23] and show distorted bond angles in the central ring.

| **4** | **5** | **6** |

These results support the suggestion that the energy difference between folded and planar thianthrene conformations is small. However, 2,3,7,8-tetramethoxy-thianthrene radical cation [22], as both the hexachloroantimonate and triiodide salts, and thianthrene radical cation [20], as its tetrachloroaluminate salt, have been shown to be planar in the solid state. The dihedral angles Θ are 180° and 174°, respectively. These results provide strong evidence for delocalization in these radical cations. A D_{2h} planar radical cation, in a deep well, was found by PM3 calculations on thianthrene [20].

The reactions of thianthrene radical cation have recently been reviewed [10]. Consequently, this discussion will emphasize the major reaction motifs and reactions reported subsequent to the previous review. However, some of the previously reviewed material will be presented here so that a complete picture may be painted.

1.1.1
Electron Transfer Reactions

Thianthrene radical cation is a good one-electron oxidant. The first reversible one-electron oxidation potential for thianthrene is +1.23 V in acetonitrile vs SCE. The reaction of thianthrene radical cation with organometallic reagents typically occurs via initial one-electron transfer. Thus reaction of phenylmagnesium chloride and phenyllithium with thianthrene radical cation produce predominantly benzene and thianthrene [25]. Labeling studies showed that the benzene is formed via phenyl radical which abstracts a hydrogen atom from the solvent. Treatment of dialkylmercury [26, 27], and tetraalkyltin [28] reagents with thianthrene radical cation also proceed via initial electron transfer. In the case of diethylmercury, ethyl radical, formed by fragmentation of the organometallic radical cation produced by electron transfer, was trapped with molecular oxygen. It should be noted that thianthrene radical cation appears to be inert to molecular oxygen in contrast to alkyl radicals. This difference in reactivity permits the selective trapping of ethyl radical in the presence of large amounts of thianthrene radical cation. The reaction of organomercury and organotin compounds with thianthrene radical cation is outlined in Eqs. (1–3). Subsequent to electron transfer as shown in Eq. (1) fragmentation occurs as shown in Eq. (2) to provide R· which then undergoes radical coupling with thianthrene radical cation to give the corresponding sulfonium salt 7 as shown in Eq. (3):

$$1 + RM \rightarrow TH + [R-M]^{+\cdot} \tag{1}$$

$$[R-M]^{+\cdot} \rightarrow R\cdot + M^+ \tag{2}$$

$$1 + R\cdot \rightarrow \quad \tag{3}$$

7

In competition with this radical coupling reaction is electron transfer between the alkyl radical and thianthrene radical cation. If the radical R· is of sufficiently low oxidation potential (and the reorganization energy permits), e.g., tert-butyl radical, then electron transfer occurs to yield the corresponding cation R^+. Evidence for electron-transfer in these reactions was also convincingly demonstrated by the use of unsymmetrical organometallic species. Kochi [29] has shown that the alkyl group that is preferentially cleaved in electron-transfer reactions is that which forms the most stable radical. Thus for R=Et, i-Pr, and t-Bu, RSnMe$_3$ undergoes selective R scission rather than Me in either inner sphere or outer sphere electron-transfer reactions. However, Me is preferentially cleaved in reactions involving electrophilic attack on the alkyl group. Selective cleavage of R over Me occurs in the reactions of RHgMe [27] and RSnMe$_3$ [28] with thianthrene radical cation. Interestingly, there is no evidence for the formation of phenyl radicals in the reaction of diphenylmercury with thianthrene radical cation [30]. Indeed treatment of **8** with thianthrene radical cation does not produce any product derived from the cyclized radical **9** expected if the corresponding phenyl radical were formed even fleetingly.

8   9

This suggests that the mechanism for the formation of arylsulfonium salt **7**, R=Ar involves the complexation mechanism [31] further discussed below, but illustrated for the first part of this reaction in Eqs. (4–6).

$$Ar_2Hg + 1 \; \rightleftharpoons \; (complex)^{\overset{+}{\cdot}} \tag{4}$$

$$(complex)^{\overset{+}{\cdot}} + 1 \; \rightleftharpoons \; (complex)^{2+} + 2 \tag{5}$$

$$(complex)^{2+} + \; \rightarrow \; 7, \; R = Ar + ArHg^+ \tag{6}$$

Thianthrene radical cation is also an excellent one-electron oxidant of iron porphyrin complexes. Such oxidation of Fe^{III}(OClO$_3$)(TPP), where TPP is meso-tetraphenylporphyrin, provides the corresponding porphyrin π-cation radical analytically pure [32]. Similar oxidation of the N-methyl porphyrin complex (N-MeTPP)Fe^{II}Cl, where N-MeTPP is N-methyl-meso-tetraphenylporphyrin, afforded [N-MeTPPFe^{III}Cl]$^+$ which was not further oxidized [33]. Thus thianthrene radical cation selectively oxidized the aromatic porphyrin ligand in one case and the metal center in the other. Ligand oxidation at a phenolic moiety has also been reported [34] on treatment of a 1,4,7-triazacyclononane appended with one or two phenol moieties ligated to Cu(II) complex with thianthrene radical cation.

Formation of π-cation radicals has also been suggested [35] in the reaction of stable enols with thianthrene radical cation and other one-electron oxidants. Thus enol **10** on treatment with thianthrene radical cation affords benzofuran derivative 11 in 87 % yield. The initial step in this reaction is suggested to be one-electron transfer forming the cation radical of 10, which has been unequivocally identified in a related system [36].

10   **11**

This cation radical rapidly deprotonates, undergoes further oxidation (or disproportionates) to the corresponding α-carbonyl cation. Cyclization of this cation and rearrangement [37] yield ultimately **11**. It should be noted that an alternative reaction path has been identified in the reaction apparently of enols with thianthrene radical cation. Ketones [38, 39] and aldehydes [40] on treatment with thianthrene radical cation form sulfonium salts **12** and thianthrene.

12

Cyclized products **13** and **14** are formed in 0.7 and 12 % yields, respectively, in the reaction of lithium dialkylamide **15** with thianthrene radical cation [41].

13   **14**   **15**   **16**

The formation of these products is cited as evidence for electron-transfer resulting in amino radical **16** which is expected to cyclize. The cyclized radical abstracts a hydrogen atom to produce **13** or preferentially is further oxidized by thianthrene radical cation to the corresponding carbocation which loses a proton to yield **14**. Redox catalysis of the Diels-Alder dimerization of 1,3-cyclohexadiene by thianthrene radical cation has been reported [42]. Although electron transfer to give the π-cation radical from the diene which then reacts with diene is conceivable, this is not believed to occur. Use of other typical one-electron oxidants with high oxidation potentials instead of thianthrene radical cation do es not give these dimers. Consequently, an inner-sphere mechanism in which there is transient bond formation between thianthrene radical cation and diene is suggested.

Thianthrene radical cation undergoes self exchange, i.e., transfer of an electron from thianthrene to its radical cation and the rate of this process measured [43]. In addition, this exchange has been studied by measurement of the equilibrium constant for the reaction shown in Eq. (7) using deuterated thianthrene. In $CF_3CO_2H/(CF_3CO)_2O/CH_2Cl_2$ at room temperature, the equilibrium constant for Eq. (7) is 0.62 ± 0.12:

$$TH + [^2H_8]TH^{\overset{+}{\cdot}} \rightleftharpoons TH^{\overset{+}{\cdot}} + [^2H_8]TH \tag{7}$$

Disporportionation, as shown in Eq. (8) involves electron transfer from one thianthrene radical to another:

$$2TH^{\overset{+}{\cdot}} \rightleftharpoons TH + TH^{2+} \tag{8}$$

This equilibrium has attracted much attention because of the mechanism first suggested for the reaction of thianthrene radical cation with water. In this reaction, the second order dependence of the rate on the concentration of $TH^{\overset{+}{\cdot}}$ was explained by disproportionation followed by nucleophilic attack on the dication. The equilibrium constant K_D for the reaction shown in Eq. (8) may be determined electrochemically. The difference in $E^{0'}$ for the two one-electron processes shown in Eqs. (9) and (10) determine K_D. Interestingly, this value depends strongly on solvent and counterion and pK_D ranges between 5–13 [46–48]:

$$TH^{\overset{+}{\cdot}} + e^- \rightleftharpoons TH \tag{9}$$

$$TH^{\overset{+}{\cdot}} \rightleftharpoons TH^{2+} + e^- \tag{10}$$

Thianthrene dication **17** is formally a fully aromatic species but evidence for electronic stabilization of these species has yet to be adduced.

However, the crystal structure for 2,3,7,8-tetramethoxythianthrene dication sheds light on the nature of its delocalization [49]. Rather than an aromatic structure the bond lengths suggest two delocalized $(MeOCCCSCCCOMe)^+$ moieties as shown in **18** joined by elongated C2–C3 and C11–C12 bonds of 1.472 and 1.453 Å which are 0.06 and 0.07 Å longer than the corresponding bonds in the parent 2,3,7,8-tetramethoxythianthrene.

Anion radicals, such as 2,5-diphenyl-1,3,4-oxadiazole anion radical, undergo electron-transfer with thianthrene radical cation resulting in chemiluminescence [10]. That is, the energetics of such radical annihilations are sufficiently great to generate an excited state which then emits light. Since the radical anion and radical cation can be generated electrochemically, such reactions gives rise to electrogenerated chemiluminescence. Similar reactions resulting in electrogenerated chemiluminescence can be carried out in aqueous solutions using sodium 9,10-diphenylanthracene-2-sulfonate and 1- or 2-thianthrene carboxylic acid [50]. Electron transfer between C_{60} anion radical and thianthrene radical cation resulting in emission at 720 nm apparently due to the triplet state of C_{60} has been reported [51, 52].

Oxidation of azo compounds by thianthrene radical cation has been reported. Thus treatment of 1,1'-azoadamantane 19 with two equivalents of thianthrene radical cation perchlorate in acetonitrile produced N-adamantylacetamide 20 in 90 % yield and thianthrene quantitatively [53].

19   20

This led to the interesting suggestion that the azo compound underwent one-electron oxidation to the then little known azoalkane radical cation. This species subsequently lost dinitrogen to form 1-adamantyl cation and 1-adamantyl radical. This latter species was then oxidized by thianthrene radical cation to 1-adamantyl cation. The 1-adamantyl cation thus formed underwent Ritter reaction to produce N-adamantyl acetamide 20.

21   22   23a, R=Me
                                                          b, R=H

Other azo compounds such as 21–23 a were found to undergo comparable reaction [54], although 23b forms adduct 24 owing to less favorable fragmentation than with 23b.

24   25

To adduce direct evidence for the formation of a radical as well as a cation on fragmentation of azoalkane radical cations, the oxidation of azoalkane 25 was studied [55].

If fragmentation of the radical cation from 25 afforded radical 26 then five-membered ring products derived from intramolecular cyclization of this radical should be formed. Treatment of 25 with thianthrene radical cation gave 27–29,

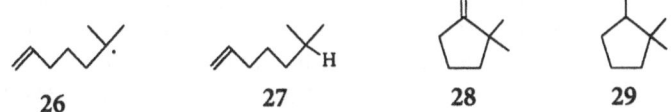

26 **27** **28** **29**

products expected from radical **26**, in a combined yield of approximately 10 %. The bulk of the products formed in this reaction were ascribable to cationic intermediates. Thus oxidation of radical 26 by thianthrene radical cation is especially fast or the azo radical cation is oxidized to a dication before decomposition. This dication then fragments into two cations averting the formation of radical 26.

Oxidation of aryl hydrazones by thianthrene radical cation have also been suggested to occur via electron-transfer and such reactions have been reviewed previously [110]. Reaction of oximes with thianthrene radical cation produces cycloaddition products [56, 57], nitriles, and carbonyl compounds. The cycloaddition products are believed to be formed via initial one-electron oxidation of the oxime.

Finally it should be noted that the radical products obtained by nucleophilic attack on thianthrene radical cation are characteristically oxidized by this radical cation in a one-electron transfer reaction. This process is presented in detail in the subsequent section on nucleophilic attack.

1.1.2
Radical Behavior

There is a scant data on the radical behavior of thianthrene radical cation [58]. As pointed out above, thianthrene radical cation couples with radicals. It shows little reactivity toward oxygen but this may be ascribed to its positive charge which should render it an electrophilic radical. It initiates the polymerization of styrene [59, 60], α-methyl styrene [59], and ethyl vinyl ether [59, 60], but not that of vinyl pyridine [61], vinyltrimethylsilane [59], methyl acrylate [59, 61], or acrylonitrile [59, 61]. These results can be explained by cationic rather than radical polymerization.

Coupling of thianthrene radical cations is another potential manifestation of radical-like behavior. Such coupling is, of course, disfavored electrostatically. Little is known about the interaction of these species but the scant information available is intriguing. Spectroscopic changes and particularly the observation of diamagnetism on increasing the concentration of thianthrene radical cation perchlorate in propionitrile led to the suggestion that this species aggregates in solution [62]. Association of two species reversibly forming a π-complex in which the spins were correlated but without covalent bond formation accounts for the observed spectroscopic changes. Although it has been suggested that some thianthrene radical salts dimerize in the solid state [4], other salts are paramagnetic with an estimated one unpaired electron per thianthrene moiety [3, 4]. The crystal structure studies of thianthrene radical cation tetrachloroaluminate [20] and 2,3,7,8-tetramethoxythianthrene radical cation triiodide salt

[23] reveal cofacial stacks of pairs of radical ions. The average intermolecular S$\cdots$S separation is 3.08 and 3.16 Å, respectively, which is longer than an S-S bond length but less than twice the van der Waals radius of sulfur (3.70 Å). In the 2,3,7,8-tetramethoxythianthrene radical cation $SbCl_6^-\cdot CH_3CN$ salt [23] the cofacial rings are equally space with the average intermolecular S$\cdots$S distance of 3.66 Å.

1.1.3
Attack by Nucleophiles

The reaction of $TH^{\ddot{+}}$ with nucleophiles has been recently reviewed [10]. Consequently, only the new developments published subsequent to that review will be presented here. However, because of the importance of the mechanistic possibilities, this section will begin with a brief discussion of these possibilities.

The mechanisms for the reaction of $TH^{\ddot{+}}$ presented here have been thoroughly discussed elsewhere [31, 63]. The disproportionation mechanism was discussed above and is shown in Eqs. (11–13). The complexation mechanism shown in Eqs. (14–16) is the generally accepted mechanism for the reaction of nucleophiles with $TH^{\ddot{+}}$.

Disproportionation Mechanism

$$2\ TH^{\ddot{+}} \rightleftharpoons TH^{2+} + TH \tag{11}$$

$$TH^{2+} + NuH \rightarrow TH^+ - \overset{+}{N}uH \tag{12}$$

$$TH^+ - \overset{+}{N}uH \rightarrow TH^+ - Nu + H^+ \tag{13}$$

The half-regeneration mechanism shown in Eqs. (17–19) is similar to the complexation mechanism except that a covalent bond is formed between $TH^{\ddot{+}}$ and the nucleophile in the half-regeneration mechanism but only a complex, devoid of a covalent bond, is formed between these species in the complexation mechanism. This raises the interesting issue of how reactive radical cations are toward nucleophiles. Low reactivity of some aromatic radical cations with nucleophiles, compared with even-electron species of the same charge, had been noted experimentally [64].

Complexation Mechanism

$$TH^{\ddot{+}}\ NuH \rightleftharpoons (TH^{\ddot{+}}\ NuH) \tag{14}$$

$$(TH^{\ddot{+}}\ NuH) + TH^{\ddot{+}} \rightleftharpoons (TH^{2+}NuH) + TH \tag{15}$$

$$(TH^{2+}NuH) \rightarrow TH^+ - Nu + H^+ \tag{16}$$

Half-Regeneration Mechanism

$$\overset{\bullet+}{TH} + \overset{+}{NuH} \rightleftharpoons (TH - \overset{+}{NuH}) \tag{17}$$

$$\overset{\bullet}{TH} - \overset{+}{NuH} + \overset{\bullet+}{TH} \rightleftharpoons (TH^+ - \overset{+}{NuH} + TH) \tag{18}$$

$$\overset{+}{TH} + \overset{+}{NuH} \rightarrow \overset{+}{TH} - NuH^+ \tag{19}$$

Furthermore, analysis of such reactions using the configuration mixing model suggested that there should be a high energy barrier for such reactions [65]. Nevertheless, it has been shown that nucleophilic attack on 9-phenylanthracene radical cation by pyridines [66] and acetate [67] is rapid (with bimolecular rate constants of 10^5–10^9 $M^{-1}s^{-1}$) with a low energy barrier. Similarly kinetic studies using basic flash photolysis techniques on the nucleophilic attack of water and imidazole on $TH^{\bullet+}$ in acetonitrile reveal bimolecular rate constants of 4.9×10^4 $M^{-1}s^{-1}$ and 3.6×10^7 $M^{-1}s^{-1}$, respectively [68,69]. A theoretical reevaluation of this issue has been published [70]. A photocatalytic system based on poly-l-histidine for oxidizing water-soluble thianthrene-2-acetic acid to a mixture of the corresponding sulfoxides based on these nucleophilic reactions of $TH^{\bullet+}$ analogs has been reported [71].

Reaction of alcohols with $TH^{\bullet+}$ has been studied [72]. The overall reaction is shown in Eq. (20):

$$ROH + 2TH^{\bullet+} \longrightarrow \quad \text{(structure 30)} \quad + TH \tag{20}$$

30

Alkoxysulfonium salt **30**, R=cyclohexyl, is isolable and has been fully characterized [73]. Decomposition of these salts in acetonitrile leads to alkenes, ethers, and *N*-alkylacetamides. Notably absent from this list of decomposition products are aldehydes and ketones which are typically formed by the decomposition of alkoxysulfonium salts **31** derived from dimethylsulfoxide [74, 75]. The lack of formation of aldehydes and ketones from **30** is consistent with the known mechanism [76–78], shown in Eq. (21), for formation of these compounds from **31**:

$$\underset{H \quad CH_3}{RR'C\overset{+}{OSCH_3}} \longrightarrow \underset{H \quad CH_2}{RR'C\overset{O}{\underset{}{C}}\overset{+}{S}CH_3} \longrightarrow \underset{+(CH_3)_2S}{RR'C=O} \tag{21}$$

31   **32**

that is, formation of a sulfonium ylide **32** followed by decomposition involving intramolecular proton-transfer. Formation of sulfonium ylides from alkoxysulfonium salts **30** is precluded and, therefore, carbonyl compounds are not formed. 2,3-Dimethyl-2,3-butanediol apparently reacts with $TH^{\bullet+}$ in a manner similar to that of monoalcohols [79]. An alkoxysulfonium salt is formed which decomposes to give pinacolone and tetramethyloxirane. However, tetraaryl-pinacols [78] undergo electron-transfer with TH⁺̇ followed by C-C bond cleavage to afford diaryl ketones after loss of a second electron.

Phenols [80], in analogy with other activated aromatic compounds, such as anisoles, react with TH⁺̇ to produce S-aryl sulfonium salts as shown in Eq. (22):

$$(22)$$

However with 2,4,6-trisubstituted phenols electron-transfer occurs followed by nucleophilic attack by nitrile solvent. Alkyl group rearrangement or loss ultimately leads to benzoxazole products.

The reaction of TH⁺̇ with nitrite and nitrate ions has been reinvestigated because of its relevance to air oxidation of TH to thianthrene sulfoxide (THO) catalyzed by NO, NO_2, or $NO^+BF_4^-$ [81, 82]. Nitrite and nitrate ions react with TH⁺̇ to generate THO without regenerating TH [81–83]. The kinetics of the reaction of nitrate ion with TH⁺̇ has been studied [84]. The reaction is second order in TH⁺̇ and first order in NO_3^-. The mechanism suggested to account for this reaction is shown in Eqs. (23–26). In the catalytic version of this reaction TH is oxidized to TH⁺̇ by $NO^+BF_4^-$, which is also known [85] as a stoichiometric reaction, concomitantly producing NO. Oxygen is known to oxidize NO to NO_2. In the presence of electron-rich aromatic compounds NO^+ is known to be generated from NO_2, thereby continuing the catalytic cyclic and enabling the use of NO or NO_2 as a catalyst. This process is accomplished as shown in Eqs. (27–29) where [TH, $NO^+]_{CT}$ is a charge transfer complex.

$$TH^{+̇}NO_3^- \rightleftharpoons T\dot{H}-ONO_2 \qquad (23)$$

$$TH^{+̇}+T\dot{H}-ONO_2 \rightarrow TH^+-ON(O^-)O^+TH \qquad (24)$$

$$TH^+-ON(O^-)O-\dot{T}H \rightarrow THO+ONO\overset{+}{T}H \qquad (25)$$

$$ONO^+TH \rightarrow THO+NO^+ \qquad (26)$$

$$2\ NO_2 \rightleftharpoons ON^+\ NO_3^- \qquad (27)$$

$$TH+ON^+NO_3^- \rightleftharpoons [TH,\ NO^+]_{CT}NO_3^- \qquad (28)$$

$$[TH,\ NO^+]_{CT} \rightarrow TH^{+̇}+NO \qquad (29)$$

Presumably, nitrate ion attacks TH⁺̇ as outlined above to form THO. Since O_2 does not oxidize TH nor react with TH⁺̇ both of these functions must be accomplished by nitrogen oxides to provide THO from TH.

The reactions of TH^{+} with other nucleophiles have been reviewed previously [10, 86–88]. As mentioned above potential nucleophiles may preferentially undergo electron-transfer with TH^{+}. Thus nucleophilic attack and electron transfer reactions compete with each other. Prediction of which of these reactions occurs is an interesting but, as yet, unresolved issued [89]. Clearly an important factor is the oxidation potential of the nucleophile. However, electron transfer may occur by an outer sphere or inner sphere mechanism [90, 91] thereby complicating the analysis of electron transfer. Future work may provide the insights necessary to fully understand this issue.

1.2
Thianthrene Analogues

Compounds **33** and **34** are analogues of TH in the sense that the central 1,4-dithiin ring is annulated to two aromatic moieties.

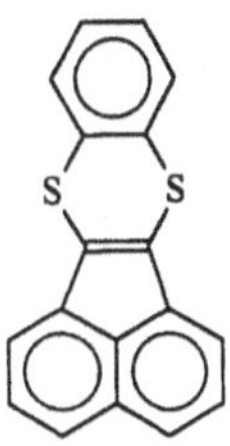

33 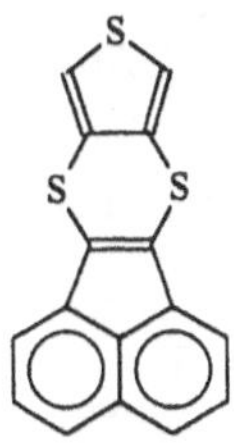   **34** 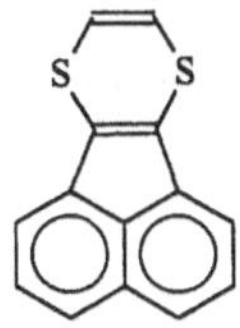   **35**

The central 1,4-dithiin ring in **33** adopts a boat conformation, as is the case for TH, with a 48° fold along the S-S vector in the solid state as shown by X-ray crystallographic analysis [92]. Both **33** and **34** undergo reversible one-electron oxidation to the corresponding radical cation with $E_{1/2} = 0.93$ and 0.77 V, respectively, vs Ag/AgCl as shown by cyclic voltammetry in acetonitrile [92]. There is also an irreversible second oxidation for both compounds at $E_{1/2}$ ca. 1.5 V.

The radical cations of benzo-1,4-dithiin **36**, 1,4-dithiin **37a** and some of their derivatives have been prepared and characterized by their EPR spectra [7, 93–96].

The average g-values for **37a** radical cation, and its substituted derivatives are 2.0077–2.0098. The ^{1}H hyperfine splittings for **36** radical cation are $a_H 2 = 3.20$ [93], 3.32 [7], $a_H 5 = 0.201$ [7], and $a_H 6 = 1.05$ [93], 1.056 [7] G. The corresponding

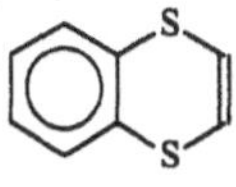

36 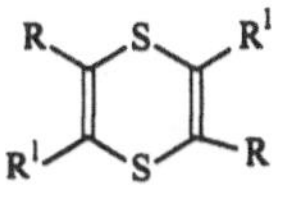   **37a, R=R^1=H**   **38**

　　　　　b, R=H, R^1=p-XC$_6$H$_4$

　　　　　c, R, R^1=SMe

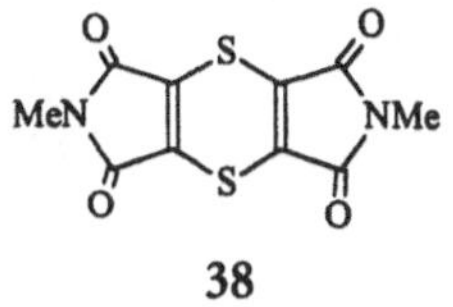

splitting for the ring hydrogens in **34a** radical cation and its substituted derivatives is 2.10–3.50 [7, 93–96]. The radical cations of TH, **33** and **34a** constitute a series in which the effect of benzo annulation on delocalization can be evaluated. The ^{33}S hyperfine splitting in these radicals provides a measure of spin localization on sulfur [97]. The reported [7] values are 9.84, 9.35, and 9.15 G for the radical cations of **37a**, **36** and TH, respectively. Consequently, delocalization of the spin follows the order $TH^{+\cdot} > 36^{+\cdot} > 37a^{+\cdot}$.

Compound **37a** and its derivatives adopt a boat conformation in the solid state as revealed by X-ray crystallographic structure studies [98–100]. However, the 1,4-dithiin ring in **38** is planar in its complex with acridine [101]. This facilitates π-π stacking of **38** and acridine although their interaction is weak. Furthermore, calculations suggest that there is little energy difference between planar and boat conformers [102]. Thus conformational analysis of 1,4-dithiin is similar to that of thianthrene.

Electrochemical oxidation of **36**, **37a**, and their substituted derivatives has been measured. The reported $E_{1/2}$ vs SCE for the first reversible oxidations to the corresponding radical cations in acetonitrile for **37a**, **36**, and TH are +0.69, +0.80, and +0.96 V, respectively [103]. All three also show a second irreversible oxidation under these conditions at +1.16, +1.34, and +1.31, respectively. Electrochemical oxidation of **39** has been reported [104].

39

In acetonitrile there is a reversible one-electron oxidation to the corresponding radical cation at $E_{1/2} = 0.68$ V vs SCE. Electrocrystallization of this material generates a dimeric radical cation $(39)_2^{+\cdot}PF_6^-$ which shows high electrical conductivity. Cyclic voltammetric studies in acetonitrile:dichloromethane (1:1) on a series of 2,5-di(*p*-substituted phenyl)-1,4-dithiins have been made [105] and the peak potentials are shown in Table 1. The first oxidation peak is reversible except for the *p*-MeO derivative which is irreversible as are the second oxidation

Table 1. Oxidation potentials of 2,5-di(*p*-substituted phenyl)-1,4-dithiins[a] **37b**

X	E_p^1, V^b	E_p^2, V^b
MeO	0.381	1.03
Me	0.458	1.13
H	0.509	1.19
Cl	0.568	1.24
NO$_2$	0.700	1.31

[a] Data from [105].
[b] In acetonitrile: dichloromethane (1:1) vs ferrocence/ferrocinium.

peaks for all of the compounds. The lifetimes of the radical cations were determined by the reversibility of the first oxidation peak. Surprisingly the order is $NO_2 > Cl \sim H > Me > MeO$ which is the reverse of the ease of oxidation judged by E_p. The $E_{1/2}$ for oxidation of **36a** in acetonitrile is 0.561 V vs Ag/0.1 N Ag^+ in CH_3CN and there is a further irreversible oxidation at 0.965 V [106]. A series of alkylthio substituted 1,4-dithiins and condensed such systems **37b, 40, 41,** and **42** have been studied by cyclic voltammetry in acetonitrile [107].

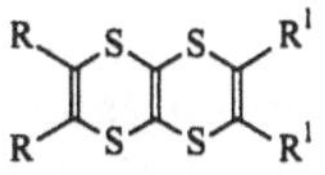

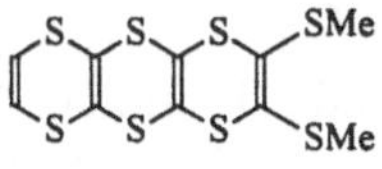

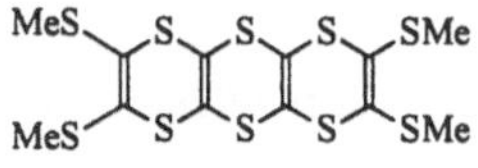

40a, R=R^1=H   **41**   **42**

b, R=SMe

R^1,R^1=SCH$_2$CH$_2$S

Each shows a reversible one-electron oxidation to the corresponding radical cation followed by a second irreversible oxidation. The reported oxidation potentials are listed in Table 2 as are the oxidation potentials for 1,4-dithiin, **37a**, itself measured under the same conditions vs the same reference electrode. Perusal of the table shows that there is no special effect on the first oxidation potential on annulating 1,4-dithiin rings although sulfur substituents significantly lower the second oxidation potential, e.g., compare **37a** and **37b**.

Table 2. Oxidation potentials of 1,4-dithiin derivatives determined by cyclic voltammetry[a]

Compound	Ep^1, V[b]	Ep^2, V[b]
37a	0.97	1.41
37c	0.95	1.10
40b	1.02	1.25
41	1.05	1.24
42	0.97	1.17

[a] Data from [107].
[b] In acetonitrile vs Ag/AgCl.

Oxidation of **40a** results in its rearrangement as shown in Eq. (30):

$$40a \longrightarrow \qquad\qquad (30)$$

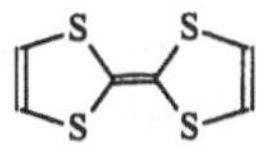

43

This rearrangement occurs via the dication of **40** but not via its radical cation. A similar rearrangement for cyclic tetrathioethylene dications is known and is discussed below.

The dimerization of the radical cations of **37b** have been reported [105]. The overall reaction is shown in Eq. (31):

$$37\,b \xrightarrow{\ -2e\ } \qquad\qquad\qquad +2H^+ \tag{31}$$

44

Kinetic analysis revealed that the radical cation of **37b** dimerizes as opposed to nucleophilic attack of **37b** on the radical cation followed by electron transfer which would be analogous to the well-known nucleophilic attack, e.g., by arenes, on TH$^+$. The dimeric products **44** were isolated in low yields (8–22 %) following preparative electrolysis. This dimerization accounts for the lifetimes of the radical cations of **37b** discussed above. Thus **37b**, X=OMe radical cation dimerizes fastest despite the lower E^0 for its parent compound.

Electrochemical oxidation of **45** has also been reported [108].

45

It shows two reversible oxidations in acetonitrile. The E^0 values for these two oxidations are 0.98 and 1.19 V vs SCE.

1.3
Diaryl Sulfide Radical Cations

In contrast to the extensive literature on the reactions of the relatively stable TH$^+$, there have been relatively few reports on diaryl sulfide radical cations. The difference in kinetic stability of TH$^+$ and Ph$_2$S$^+$ is illustrated by the reversibility of the one-electron electrochemical oxidation of TH in acetonitrile but irreversible oxidation of Ph$_2$S in this solvent [109]. In both cases the corresponding radical cation is formed but that from Ph$_2$S reacts with nucleophiles present in the system before it can be reduced back to the starting sulfide. The reported E_p for the irreversible oxidation of Ph$_2$S in acetonitrile is 1.53 V vs SCE [110] and 1.24 V vs Ag/0.1 mol/l AgNO$_3$ [111]. However, dianisyl sulfide radical cation is more stable than Ph$_2$S$^+$. A reversible or nearly reversible oxidation of dianisyl sulfide in acetonitrile is found by polarography although not by sweep voltammetry [112]. The half-life for this species is estimated as 0.7 s at 10^{-3} mol/l in acetonitrile at room temperature. The half-wave oxidation potentials of substituted diphenyl sulfides have been correlated with Hammett σ_0-values [113]. Similarly the rate of TiO$_2$ mediated photooxidation of substituted diphenyl sulfides has been found to correlate with σ^+ values giving $\rho = -0.7$ [114]. This oxidation is believed to occur via electron-transfer from the adsorbed sulfide to a photochemically generated hole in the TiO$_2$ forming a sulfur radical cation. The small negative ρ value for this oxidation is consistent with positive charge localization on sulfur in the rate determining step. Alternatively it could be due to the pre-photocatalytic step involving the equilibrium adsorption of the sulfide with TiO_2. Although Ph_2S oxidation in acetonitrile is electrochemically irreversible, its reversible oxidation potential has been determined in aqueous solutions using pulse radiolysis techniques. For the reaction shown in Eq. (32) $E^0 = 1.53$ V vs NHE has been found [115]:

$$Ph_2S - e^- \rightleftharpoons Ph_2S^{\cdot +} \tag{32}$$

Diphenyl sulfide radical cation has been reported [116] to absorb in the visible with $\lambda_m = 780$ nm for the species formed by gamma radiolysis of Ph_2S in 1,2-dichloroethane at 77 K. Laser flash photolysis of **46** in acetonitrile provides $Ph_2S^{\cdot +}$, which shows absorption bands at 330 and 750 nm [117].

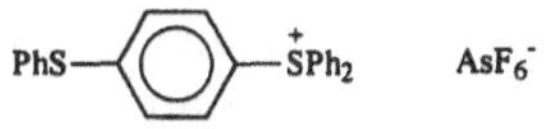

$$PhS-\langle\bigcirc\rangle-\overset{+}{S}Ph_2 \quad AsF_6^-$$

46

For this species produced under pulse radiolysis conditions in water, absorption peaks at 360 and 750 nm have been observed [115]. While $Ph_2S^{\cdot +}$ does not show any propensity to aggregate with Ph_2S in aqueous solution [115], it reacts with this species in acetonitrile and this fast reaction accounts for the irreversibility of the electrochemical one-electron oxidation of Ph_2S mentioned above. The product of this reaction is **46** [115, 118], which is suggested to form via a dispro-portionation mechanism [119] involving the corresponding dication Ph_2S^{2+} which is analogous to the disproportion mechanism illustrated above for $TH^{\cdot +}$.

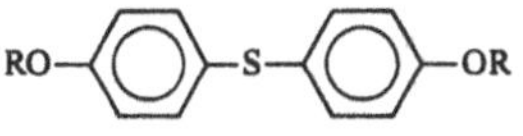

$$RO-\langle\bigcirc\rangle-S-\langle\bigcirc\rangle-OR$$

47a, R=H
b, R=Me

$$R-\langle\bigcirc\rangle-S-\langle\bigcirc\rangle-R$$

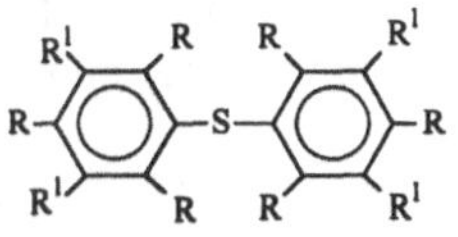

48a, R=Me, R¹=H
b, R=R¹=Me

Insight into the structure of diaryl sulfide radical cations is obtained by analysis of the EPR spectra of the radical cations obtained from **47a,b** and their alkyl substituted analogues [120–122] as well as from **48a** and **b** [95]. The hyperfine splitting values are 0.95–1.625 G for the ortho-hydrogen atoms which is comparable to that of H(2,3,7,8) in $TH^{\cdot +}$ and shows the delocalization of the spin over both rings. The relatively large g-factors of 2.0063–2.0076 are only slightly lower than that for $TH^{\cdot +}$ indicating substantial spin density on sulfur. Ab initio calculations at the STO-3G* level with geometry optimization for $Ph_2S^{\cdot +}$ suggest that there is delocalization resulting in small bond length changes compared with Ph_2S leading toward a quinoid-type structure [123]. More marked geometry changes are found in the corresponding dication [123].

Triarylsulfonium salts are photoinitiators of cationic polymerization. It has been suggested [124, 125] that they undergo homolysis on irradiation to generate diaryl sulfide radical cations and aryl radicals. However, this suggestion has proved controversial. Evidence has been presented that direct irradiation of triarylsulfonium salts proceeds via the singlet excited state to give mainly heterolysis to diarylsulfide and aryl cation [126, 127]. Although direct irradiation of **46** [127], which, unlike direct irradiation of the other triaryl sulfonium salts, decomposes via the triplet state, and triplet sensitized irradiation of triarylsulfonium salts, result in homolysis to afford aryl radicals and diaryl sulfide radical cations [127, 128]. Indeed laser flash photolysis of $Ph_3S^+SbF_6^-$ in acetonitrile in the presence of triplet sensitized acetone afforded Ph_2S^+ as identified by broad absorption at 760 nm [129].

Electron-transfer reactions of Ph_2S^+ with aromatic compounds have been reported [130]. For those aromatic compounds whose ionization potential is less than 780 $kJmol^{-1}$, the rates of electron-transfer are encounter controlled. For compounds with ionization potentials above this value the rate of electron-transfer decreases with increasing ionization potential. Diaryl sulfide radical cations also undergo nucleophilic attack [117, 119] by arenes, Ar_2S (from which Ar_2S^+ is formed), methanol, and water in reactions suggested to involve Ar_2S^{2+} formed by disproportionation of Ar_2S^+ as already mentioned. Table 3 lists the reported rate constants.

Nucleophilic attack by superoxide anion radical on Ph_2S^+ is apparently involved in the photooxidation of Ph_2S to Ph_2SO sensitized by 1,4-dimethoxynaphthalene (DMN) in polar solvents [131]. In this reaction DMN absorbs the light forming an excited state which transfers an electron to O_2 forming DMN^+ and O_2^-. Electron transfer from Ph_2S to DMN^+ results in the formation of Ph_2S^+ which reacts with O_2^- to produce $Ph_2S^+OO^-$. This species, or its subsequently formed isomer thiadioxirane [132], reacts with Ph_2S to yield two molecules of Ph_2SO in a reaction well-precedented by singlet O_2 oxidation of sulfides [133–136]. An alternative mechanism involving singlet oxidation of Ph_2S is

Table 3. Rates of reaction of Ar_2S^+ with nucleophiles

Nucleophile	k, $Lmol^{-1}$ s^{-1}			
	Ar=Ph	$4-MeC_6H_4$	$4-MeOC_6H_4$	$2,4,6-Me_3C_6H_2$
Ar_2S	1.1×10^{7a}			
	6×10^{7b}	5×10^{6b}	nr[b]	nr[b]
$1,3,5-Me_3C_6H_3$	5×10^{5b}	5×10^{6b}	40[b]	100[b]
PhOMe	–	–	240[b]	–
MeOH	5.4×10^{3a}	–	–	–
H_2O	2.3×10^{3a}			–
	10^{7b}	10^{6b}	80[b]	

[a] Ref. [117].
[b] Ref. [119].

ruled out because Ph$_2$S is relatively unreactive toward singlet O$_2$ [132–136] and this photooxidation is rapid.

1.4
Alkyl Aryl Sulfides

Alkyl aryl sulfide radical cations have attracted some interest. A number of such radical cations have been made by reacting the sulfide with aluminum chloride in nitromethane or dichloromethane and characterized by EPR spectroscopy [137–141]. The reported g$_{av}$ values are 2.0069–2.0087 and the hyperfine splitting constants, a$_H$, for the aromatic hydrogens are 0.73–5.44G and for the α-aliphatic hydrogens 1.03–2.56 G. π-Delocalization is evident but comparison of such sulfur radical cations with the corresponding oxygen analogues reveals the greater delocalization in the oxygen than sulfur radical cations [138].

In addition to π-delocalization of alkyl aryl sulfide radical cations, other possibilities exist for stabilization of the radical cation if there are electron-rich groups with unpaired electrons ortho to the sulfur atom. Because of the importance of these possibilities, they will be discussed in detail. The radical cation of **49a** may prefer a two-center, three-electron (2c, 3e) bond between the two sulfur atoms as illustrated in **50** rather than π-delocalization.

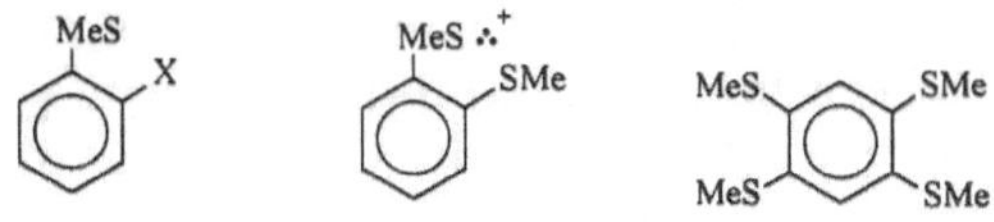

49a, X=SMe   **50**   **51**
   b, X=CO$_3$t-Bu

Such bonding can be envisioned simply as follows. A p-type orbital on each of the sulfur atoms is arranged for σ-type overlap in the plane of the aromatic ring, i.e., orthogonal to the π-system. This results in formation of σ and σ* molecular orbitals. There are three electrons to be accommodated in these MOs resulting in a σ^{2}σ*^{1} electronic ground state. This bonding scheme is depicted in Scheme 1 and gives rise to a formal bond order of 1/2.

Scheme 1

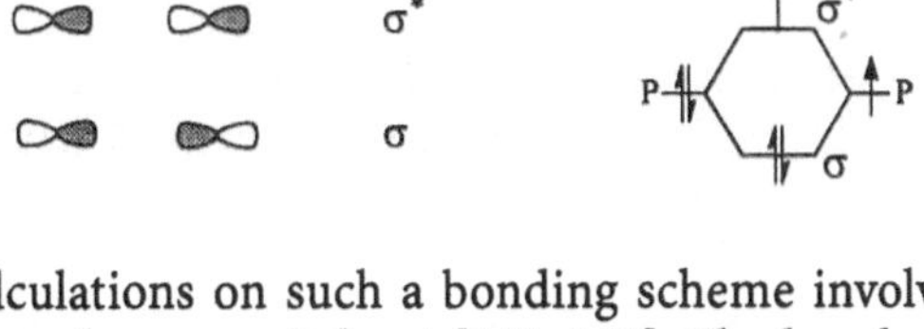

Theoretical calculations on such a bonding scheme involving a variety of different atoms have been carried out [142, 143]. The bond strength for such bonds depends on the difference in ionization potential (ΔIP) between the two centers. The smaller this difference, the stronger the bond. Thus S,S 2c, 3e bonds are predicted to be stronger than S,O 2c, 3e bonds. In addition, as the overlap integral between the two sulfur atoms increases beyond 1/3, the bond strength decreases [144]. This decrease results from the steeper increase in energy of the

σ* orbital than decrease in energy of the σ orbital with increasing overlap. Intermolecular 2c, 3e S,S bond formation is very important in dialkyl sulfides, as is discussed in detail below, but it is not important in alkyl aryl sulfides [145] or diphenyl sulfide [115]. However the intramolecular case is very interesting. Based on the EPR spectroscopic parameters reported for the radical cation of 51, i.e., g_{av}=2.0077, a_H(ArH)=0.73 G and a_H(Me)=2.59 G, it is concluded that it is a π-delocalized species. However, oxidation of 52 gives the corresponding radical cation with a five line EPR spectrum with g_{av}=2.0133 and a_H=4.75 G [146].

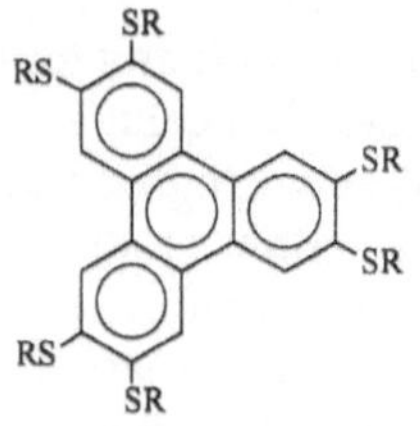

52, R=CH₃(CH₂)₅ 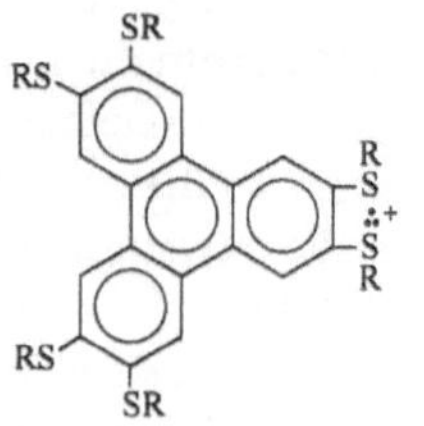   53, R=CH₃(CH₂)₅

These data demonstrate that a π-delocalized radical cation is not formed but rather one with a 2c, 3e bond as shown in 53. Furthermore, the visible absorption spectrum of this species, λ_m=484 nm, is that expected from such a species.

Another bonding scheme has been suggested for that in radical 54 produced by decomposition of perester 49b [147–151].

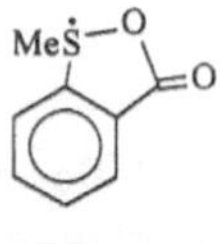

54 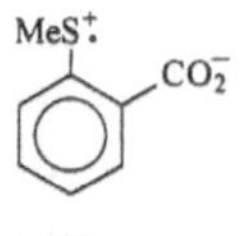   55

This is a 9-S-3 [147] species like 50 and can be regarded as being derived by interaction of the electron-rich carboxylate moiety with the sulfur radical cation in 55. Analysis of the EPR spectrum of this species, which shows splitting (1.5 G) by only one of the aromatic hydrogens (H6), using semiempirical computational methods, led to the suggestion that there is a 3c, 3e bond involving linearly arranged O-S-C (Me). Such a bonding scheme has been described for σ-sulfuranyl (O-S-O) radicals [149] and is depicted in Scheme 2. In this bonding arrangement for 54 the odd electron is in the σ-system.

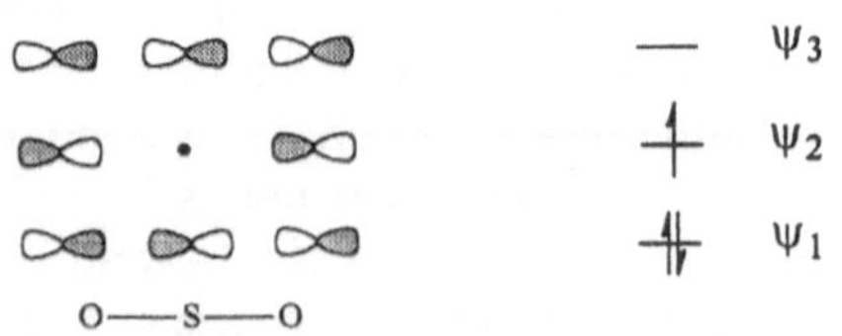

Scheme 2   O——S——O

An alternative bonding scheme is a π-delocalized radical with a linear 3c, 4e bond involving O-S-O. Such π-radicals have been suggested for radicals **56a,b** based on analysis of their spectra obtained using EPR and ESE techniques [152].

56a, R=CF$_3$, R^1=H
b, R=CF$_3$, R^1=t-Bu

A simplified view of this bonding scheme is shown in **57**, in which the odd electron is in a p-type orbital on sulfur oriented for overlap with the aromatic π-system.

The radical cations of thioanisole, **58a–e**, show two strong absorptions between 300–350 and 500–600 nm in aqueous solution [145].

58a, R=Me, R$_m$=R$_p$=H   **59**   **60**   **61**
  b, R=R$_p$=Me, R$_m$=H
  c, R=PhCH$_2$, R$_m$=R$_p$=H
  d, R=PhCH$_2$, R$_m$=H, R$_p$=CH$_2$SO$_3$K
  e, R=p-MeOC$_6$H$_4$CH$_2$, R$_m$=H, R$_p$=CH$_2$SO$_3$K
  f, R=Me, R$_m$=H, R$_p$=OMe
  g, R=MeOCH$_2$, R$_m$=R$_p$=H
  h, R=MeOCH$_2$, R$_m$=H, R$_p$=OMe

Electrochemical oxidation of alkyl aryl sulfides **58** has been studied in acetonitrile using cyclic voltammetry [110, 111, 153, 154]. The first oxidation potentials are listed in Table 4. Other oxidations are observed at higher potentials but, in general, the first oxidation peak corresponds to removal of an electron from sulfur to generate the corresponding radical cation. The oxidation potentials for the series of p-substituted thioanisoles: p-CF$_3$, p-Me, p-OMe, p-NMe$_2$, and thioanisole itself listed in the third column – linearly correlate with σ_p^+ [153]. Linear correlation of the oxidation potentials with σ^+ was also reported [110] for the substituted-phenyl benzyl sulfides whose potentials are listed in the second column of Table 4. A large negative r value of –3.3 was found for this series. The peak potentials for oxidation of **59** and **60** in acetonitrile have been reported [111] as 1.28 and 0.85 V, respectively, vs Ag/0.1 mol/l AgNO$_3$ in CH$_3$CN. In the presence of activated neutral alumina added to ensure anhydrous conditions p-MeO, p-Br, p-Cl, p-CN, and p-NO$_2$ thioanisole but not thioanisole; **58**, R=Et, R$_m$=R$_p$=H; **58**, R=i-Pr, R$_m$=R$_p$=H; **58**, R=PhCH$_2$, R$_m$=R$_p$=H; **59** or **60** showed

Table 4. Oxidation potentials[a] of alkyl aryl sulfides **58** in acetonitrile

R	R_m	R_p	$E_{1/2}$, V[b] vs Ag/0.01 mol l^{-1} Ag$^+$	E_p, V[c] vs SCE	E_p, V[d] vs Ag/AgCl	E_p, V[e] vs Ag/0.1 mol l^{-1} AgNO$_3$, CH$_3$CN
Me	H	H	1.10	1.47	1.56	1.12
Et	H	H	–	–	–	1.16
i-Pr	H	H	–	–	–	1.20
PhCH$_2$	H	H	–	1.55	–	1.20
Me	H	CF$_3$	–	–	1.80	–
Me	H	Br	–	–	–	1.14
Me	H	Cl	–	–	–	1.17
Me	H	CN	–	–	–	1.37
Me	H	NO$_2$	–	–	–	1.44
Me	H	Me	0.93	–	1.48	1.05
Et	H	Me	0.98	–	–	–
i-Pr	H	Me	1.05	–	–	–
t-Bu	H	Me	1.14	–	–	–
Me	Me	H	1.02	–	–	–
Me	H	OMe	–	–	1.43	0.86
Me	H	NMe$_2$	–	–	0.88	–
Et	Me	H	1.03	–	–	–
i-Pr	Me	H	1.08	–	–	–
t-Bu	Me	H	1.19	–	–	–
PhCH$_2$	Cl	H	–	1.67	–	–
PhCH$_2$	H	Cl	–	1.58	–	–
PhCH$_2$	H	Me	–	1.43	–	–

[a] First oxidation potential.
[b] [153]; [c] [110]; [d] [154]; [e] [111].

reversible oxidation to the corresponding radical cations in acetonitrile [111]. A further irreversible oxidation at higher potentials to the corresponding dications is also observed. Evidence is presented for the reversible dimerization of these radical cations. The structure of the dimers is surmised to be **61**. A number of poly(alkylthio) benzene, naphthalene, and pyrene derivatives have generally been found to undergo reversible one-electron oxidation to radical cations. The peak potentials for **62a**, **62b**, **63a**, and **63b** in acetonitrile are 1.10, 0.99, 1.17, and 1.15 V, respectively, vs SCE [155, 156]. The $E_{1/2}$ values for **64a**, **64b**, and **65** in dichloromethane are 1.21 [157], 1.18 [158], and 0.82 V [157], respectively, vs SCE. In benzonitrile containing 5 % trifluoroacetic acid anhydride the $E_{1/2}$ for **64a,b** are both 1.05 V vs SCE [159]. The irreversible peak potential for **66** in acetonitrile is 0.70 V vs Ag/0.1 M AgNO$_3$, CH$_3$CN [160].

62a, X=S
   b, X=O

63a, X=S
   b, X=O

64a, W=Y=OMe, X=Z=SMe
   b, W=X=Y=Z=SMe
   c, W=Z=SMe, X=Y=OMe

This peak potential also fits the linear correlation established between polarographic oxidation half-wave potentials for substituted aromatic compounds and their ionization potentials [161]. It should be noted that the radical cation of **66** may be either a π-delocalized or S,S 2c, 3e σ^* species. Compound **67** shows a reversible first oxidation potential in acetonitrile with $E_{1/2}$=0.87 V vs SCE [162]. Perylene derivative **68** exhibits a first oxidation peak in nitrobenzene at 0.65 V vs Ag/AgCl [163]. Because of the good thermal stability and high electrical conductivity of the radical cation of **68** [164], functionalized analogues have also been prepared and studied [165]. Poly(alkylthio)pyracyclene has been studied [166].

65   66   67   68

Cyclic voltammetric studies of **69** in benzonitrile show two reversible oxidation waves at $E_{1/2}$ 0.43 and 0.75 V vs SCE.

69

Radical cation salts of **69** were prepared by electrooxidation. The stoichiometries of two of these salts which formed crystals suitable for X-ray crystallographic analysis are $(69)_3(PF_6)_2$ and $(69)_2I_3$. The crystal structure determination show that these species consist of segregated stacks of donors and acceptors.

In addition to generating alkyl aryl sulfide radical cations electrochemically or by treatment of sulfides with $AlCl_3$ in nitromethane or dichloromethane as discussed above, they have been formed by a variety of other methods. One-electron oxidation with manganese(III) acetate, ammonium cerium(IV) nitrate, chromium(VI), or $K_5[Co^{III}W_{12}O_{40}]$ to give radical cations has been reported. Photochemical electron transfer methods and pulse radiolysis techniques have also been effectively used to produce radical cations. The radical cations so generated undergo a variety of reactions including sulfoxide formation, sulfonium salt formation, αCH deprotonation, αCSn destannylation, C-S and C-C bond cleavage, and electron transfer. Since the reactions observed depend on the conditions in which the sulfur radical cation is generated, discussion of the formation of the radical cations and their reactions are combined and presented below.

In wet acetonitrile, constant current electrolysis of thioanisole produced the corresponding sulfoxide in 74% yield [167]. Phenyl benzyl sulfide yielded the corresponding sulfoxide as well under these conditions but cleavage of the benzyl group occurred, also affording diphenyl disulfide and products from benzyl cation. Cleavage of the C-S bond can be further favored by stabilization of the carbocation formed from such cleavage. Thus phenyl triphenylmethyl sulfide formed no sulfoxide under these conditions, only diphenyl disulfide and triphenylcarbinol. Under rigorously anhydrous conditions which suppress sulfoxide formation controlled potential electrolysis of thioanisole and substituted thioanisoles give sulfonium salts along with presumably some (alkyl) C-S cleavage [111]. Such oxidation of **58b** yielded **70** as shown by X-ray crystallographic structural analysis.

70   **71**

Similarly, oxidation of **58f** gave **71** whose structure was unequivocally established by X-ray crystallographic structure studies. As already mentioned, electrochemical studies revealed that the radical cation of **58f** undergoes reversible dimerization to **61** where $Ar = p\text{-}MeOC_6H_4$. A peak assigned to the two-electron oxidation of **61**, $Ar = p\text{-}MeOC_6H_4$ results in the formation of **71**. In addition the irreversibility of the oxidation to the dication is ascribed to the rapid reaction of the dication with neutral sulfide forming the sulfonium salt **71**. It has been reported [168] that electrochemical oxidation of secondary alcohols gives the corresponding ketones in the presence of thioanisole as mediator and CF_3CH_2OH as solvent. The mechanism for this reaction involves oxidation of thioanisole followed by nucleophilic attack by the alcohol on the electrophilic sulfur to generate, in a two-electron process, an S-phenyl, S-methyl S-alkoxysulfonium salt which forms ketone in the same way as shown in Eq. (21) for S,S-dimethyl S-alkoxysulfonium salts. The nucleophilic alcohol may attack the radical cation or dication of thioanisole in this case. Anodic oxidation of $XC_6H_4CH_2SC_6H_4Y$ has been studied under a variety of conditions to evaluate the competition between α C-H deprotonation, (alkyl)C-S bond cleavage and attack on sulfur of the corresponding radical cation [169]. In acetic acid containing sodium acetate, which acts as a base, deprotonation of the cation radical is favored. It should be noted that the radical cations are strong acids, e.g., the pK_a of the cation radical of **58c** is estimated to be −3. In acetic acid containing lithium nitrate, sulfoxide formation is favored on anodic oxidation owing to attack by nitrate anion on the sulfur radical cation, a reaction analogous to that discussed above for TH⁺, but α C-H deprotonation also occurs. In acetic acid or acetonitrile containing lithium perchlorate both attack at S as well as deprotonation occur on oxidation. Apparently perchlorate anion acts as a nucleophile and undergoes reduction as evidenced by the formation of benzyl chlorides. A cation stabilizing group in the benzylic moiety favors C-S bond cleavage to yield a stabilized benzylic cation while an electron-withdrawing nitro group in this moiety disfavors this process but promotes deprotonation. Electron withdrawing groups attached to the α-carbon of phenylthioethers promotes α-deprotonation of the corresponding sulfur radical cation. For example, anodic oxidation of $PhSCH_2CF_3$ in the presence of MeOH, AcOH, or F^- generates $PhSCH(X)CF_3$ where $X = MeO$, AcO, and F, respectively, in good yields [170]. In these reactions, α-deprotonation forms a carbon radical which is oxidized to the corresponding phenylthio carbocation. Nucleophilic attack on this carbocation results in the observed products. Cleavage of (alkyl) C-S bonds is favored by a methoxy group directly attached to the carbon undergoing C-S cleavage. Thus anodic oxidation of PhSCH(OMe)R in NaOAc, AcOH affords RCH(OMe)OAc in good yields [171]. In addition, **58g** undergoes oxidation, with a potential of 1.40 V vs Ag/AgCl determined by rotating disk voltammetry, to form $MeOCH_2$ [172]. The oxidation potential of the sulfide may be rendered less positive by appending electron-releasing MeO groups to the phenyl ring attached to sulfur as seen above in Table 4. Thus **58h** and methoxymethyl (2,4-dimethoxyphenyl) sulfide have oxidation potentials of 1.18 and 1.00 V, respectively, vs Ag/AgCl. The carbocations produced by these oxidations may be trapped intermolecularly with allyltrimethylsilane and intramolecularly with alkene moieties. Thus oxidation of **72a** in the presence of allyltrimethylsilane provides **72b** in 46–70% yield as determined by ^1H NMR spectroscopy and oxidation of **73** with n-Bu_4NBF_4 as supporting electrolyte gives **74** as a mixture of diastereomers in 87% yield.

$C_{10}H_{21}CH(OMe)R$

72a, R=SAr
b, R=CH$_2$CH=CH$_2$

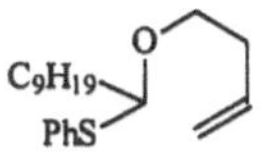

73

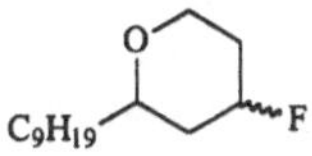

74

Of particular interest are anodic oxidations of 1-arylthioglycosides **75**, whose oxidation potentials in CH_3CN [173] are 1.85, 1.60, 1.62, and 1.64 V vs Ag/AgCl for **75**, R=Ac, R'=H, β-anomer; **75**, R=R'=H, β-anomer; **75**, R=H, R'=Me, β-anomer; **75**, R=H, R'=Me, α-anomer, respectively, which produce a glycosyl cation on C-S cleavage [173–175].

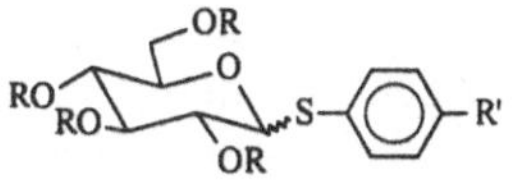

75

This species can react with a sugar alcohol to form a disaccharide. The ready availability and stability of **75** and this selective and mild glycosylation procedure render this method synthetically attractive. Anodic oxidation of diaryl dithioacetals results in S-S bond formation as well as C-S bond cleavage as shown in Eq. (33) [176, 177]:

$$(ArS)_2CH_2 \xrightarrow[2Nu^-]{-2e^-} ArSSAr + CH_2Nu_2 \qquad (33)$$

$$(ArS)_2 CH_2 \xrightarrow{-e^-} ArS\cdot + Ar\overset{+}{S}CH_2$$

$$Ar\overset{+}{S}CH_2 + Nu^- \longrightarrow ArSCH_2Nu \xrightarrow{-e^-} ArS\cdot + \overset{+}{C}H_2Nu$$

$$\overset{+}{C}H_2Nu + Nu^- \longrightarrow CH_2Nu_2$$

Scheme 3 $$2ArS\cdot \longrightarrow ArSSAr$$

$$(ArS)_2CH_2 \xrightarrow{-2e^-} \underset{Ar\overset{+}{S}\text{---}S\overset{+}{Ar}}{\overset{CH_2}{\diagup \diagdown}} \xrightarrow{Nu^-} \underset{Ar\overset{+}{S}\text{---}SAr}{\overset{CH_2Nu}{|}} \xrightarrow{Nu^-} CH_2Nu_2 + ArSSAr$$

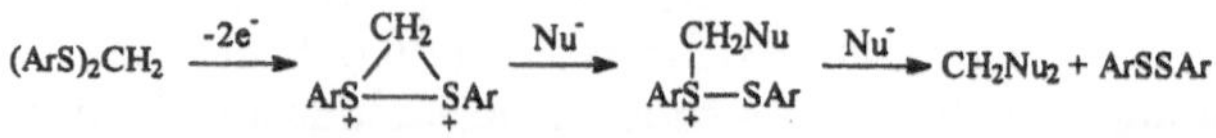

Scheme 4

Two mechanisms have been proposed for this reaction and are shown in Schemes 3 and 4. In the first [176] C-S bond cleavage precedes S-S bond formation but the reverse occurs in the second [177].

Detailed electrochemical studies and the detection of crossover disulfide products in mixed electrolyses support the first mechanism [178]. Cleavage of an α C-Si bond on oxidation of α-silyl sulfides to give phenylthio carbocations as shown in Eq. (34) M=SiMe$_3$ is known [179–182]:

$$PhSCH(M)R \xrightarrow{OX} PhS\overset{+}{C}HR \xrightarrow{Nu^-} PhSCH(Nu)R \tag{34}$$
$$\mathbf{76}$$

It is suggested that the C-Si bond of the sulfur radical cation cleaves with nucleophilic assistance to form the corresponding carbon radical which undergoes one-electron oxidation to the carbocation. Alcohols and carboxylic acids have been used as nucleophiles in these reactions. α-Silyl groups modestly lower the oxidation potentials of phenylthioethers by about 100 mV as seen in Table 5. Cleavage of an α C-Sn bond on oxidation of α-stannyl sulfides to give phenylthio carbocations as shown in Eq. (34), M=SnBu$_3$, R=C$_8$H$_{17}$ is known [186, 187] and is synthetically useful. The mechanism suggested for C-Sn cleavage is loss of Bu$_3$Sn$\cdot$ from the sulfur radical cation to directly yield the phenylthio carbocation. Such anodic oxidation in the presence of allyl trimethylsilane [186], trimethylsilyl cyanide [187] (in THF with Bu$_4$NBF$_4$ as supporting electrolyte), or the O-trimethylsilyl enol ether of cyclohexanone [186] produced **76**, Nu=CH$_2$CH=CH$_2$; **76**, Nu=NC; and **76**, Nu=2-cyclohexanone in 63, 53, and 75% yield, respectively. It should be noted that α-stannylation lowers the oxidation potential of sulfides as will be discussed further below. The oxidation potential of PhSCH(SnBu$_3$)C$_8$H$_{17}$ in MeCN is E$_{1/2}$=0.74 V vs Ag/AgCl [186].

Oxidation of alkyl aryl sulfides with manganese(III) acetate [153], ammonium cerium(IV) nitrate [110], and chromium(VI) [188] is suggested to first produce the corresponding radical cation because the rate of oxidation is linearly correlated with the corresponding electrochemical oxidation potentials. With manganese(III) acetate [153], methyl aryl sulfides produce the corresponding acetoxymethyl aryl sulfide **58**, R=CH$_2$OAc, R$_m$, R$_p$ are various substituents. With **58**, R=t-Bu, R$_m$=H, R$_p$=CH$_3$ such oxidation results in substitution at the methyl group producing **58**, R=t-Bu, R$_m$=H, R$_p$=CH$_2$OAc and **58**, R=t-Bu, R$_m$=H, R$_p$=CHO. Oxidation of alkyl aryl sulfides with ammonium cerium(IV) nitrate [110] affords the corresponding sulfoxide. Evidence is presented showing that the sulfur radical cation reacts with nitrate anion to produce the sulfoxide. Chro-

Table 5. Oxidation potentials of α-silyl sulfides and reference compounds

Compound	$E_{p/2}$, V[a] vs SCE	E_p, V[b] vs Ag/Sat Ag$^+$	E_p, V[c] vs Ag/AgCl	$E_{1/2}$, V[d] vs Ag/AgCl
PhSCH$_3$	–	1.05	–	–
PhS(CH$_2$)$_7$CH$_3$	–	–	1.20	–
PhSCH$_2$SiMe$_3$	1.15	0.92	1.10	1.29
PhSCH$_2$SiMe$_2$OCH(CH$_3$)C$_6$H$_{13}$	–	–	–	1.25
PhSCH$_2$SiMe$_2$C$_{10}$H$_{21}$	–	–	–	1.29
PhSCH$_2$SiMe$_2$F	–	–	–	1.39
PhSCH(SiMe$_3$)(CH$_2$)$_7$CH$_3$	–	–	1.10	–
PhSCH$_2$CH$_2$SiMe$_3$	1.26	–	–	–
PhS(CH$_2$)$_3$SiMe$_3$	1.28	–	–	–
PhSCH(SiMe$_3$)$_2$	–	–	–	1.18
PhSCH(SiMe$_2$F)$_2$	–	–	–	1.41
PhSC(SiMe$_3$)$_2$(CH$_2$)$_7$CH$_3$	–	–	1.10	–
(PhS)$_2$C(SiMe$_3$)(CH$_2$)$_7$CH$_3$	–	–	1.20	–

[a] [183]. [b] In CH$_3$CN [181]. [c] In CH$_3$CN [184]. [d] In CH$_3$CN [185].

mium(VI) oxidation [188] of alkyl aryl sulfides yields the corresponding sulfoxides. Nucleophilic attack by silyl enol ethers on sulfur on Ce(IV) oxidation of phenyl allyl sulfides has been reported [189]. The ultimate products from such reactions formed in 66–79% yield when R''=Ph, CH=CH$_2$, or C$\equiv$Cn-Bu result from [2, 3] sigmatropic rearrangement of the intermediary sulfonium ylides as shown in Eq. (35):

$$\text{PhSCH}_2\text{C(R)=CHR'} + \quad \xrightarrow[\text{-H}^+]{-2e^-} \quad \xrightarrow{\quad} \tag{35}$$

$$\overset{O}{\overset{\|}{\text{R''CCHR'C(R)=CH}_2}}$$

It has been suggested that K$_5$[Co^{III}W$_{12}$O$_{40}$] is an outer sphere one-electron oxidizing agent [190, 191]. This reagent has been used to oxidize sulfide 77 [192]. The products of this oxidation in aqueous acid are acetophenone:

$$\text{PhCH(Me)SPh} \xrightarrow{-e^-} 77^{\ddot+} \xrightarrow{\text{NuH}} \text{PhCH(Me)Nu} + \text{PhS}\cdot + \text{H}^+ \tag{36}$$
$$\quad\quad 77$$

which results from αC-H deprotonation, and 1-phenylethyl acetate and 1-phenylethanol, which result from C-S bond cleavage. Using a thermodynamic cycle, ΔH for C-S cleavage of a series of (ArSR)$^{\ddot+}$ in which R$^+$ is relatively stable has been found to be –6 to –24 kcal/mol [193]. To address the question of the mechanism of nucleophilic substitution on (77)$^{\ddot+}$ as outlined in Eq. (36) optically pure 77 was oxidized. The substitution reaction occurred predominantly with racemization and a small amount of inversion. This result argues for a unimolecular mechanism analogous to an S_N1 mechanism for this reaction. In this regard it is noteworthy that the radical cations of the stereoisomers of **78** undergo nucleophilic cleavage of a C-C bond with complete inversion of configuration via a mechanism analogous to the S_N2 mechanism [194].

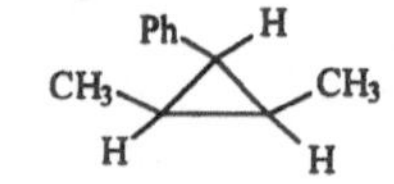

78

Inversion of configuration was also observed [195] in the C-C bond cleavage in sulfur radical cation **79** as shown in Eq. (37):

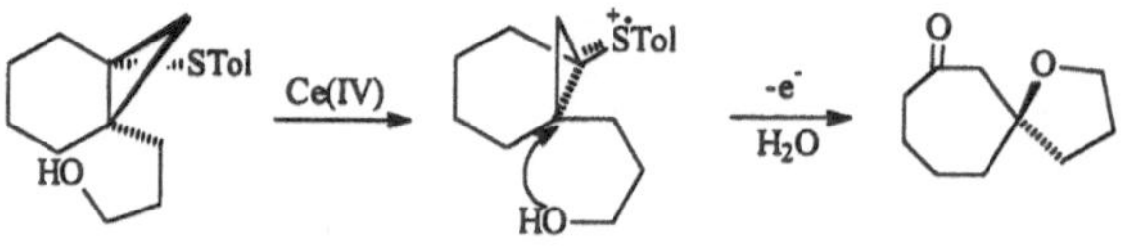

$$(37)$$

79

Photooxidation of Ph_2S to Ph_2SO sensitized by DMN in polar solvents has been discussed above. Use of this method for photooxidation of allyl phenyl sulfide has also been reported [131]. In addition, photooxidation of benzyl phenyl sulfides **80** in MeCN using 9,10-dicyanoanthracene (DCA) as sensitizer has also been reported [196].

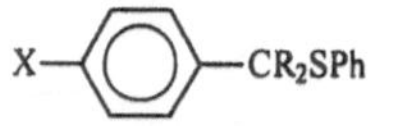

80a, R = X = H
   b, R = H, X = OMe
   c, R = Me, X = H

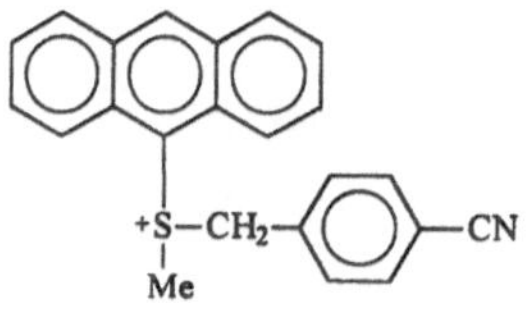

81

The reaction sequence is similar to that outlined for DMN sensitized photooxidation except that singlet DCA is the electron acceptor. It reacts with **80** to form the corresponding sulfur radical cation and $DCA^{\bar{\cdot}}$. The $DCA^{\bar{\cdot}}$ transfers an electron to O_2 forming $O_2^{\bar{\cdot}}$ and DCA. Reaction of $O_2^{\bar{\cdot}}$ with the sulfur radical cation ultimately results in the formation of the corresponding sulfoxide as before. However, the sulfur radical cations in this case also undergo α C-H deprotonation and C-S bond cleavage in addition to, or in preference to, reaction with $O_2^{\bar{\cdot}}$. Thus DCA sensitized photooxidation of **80a** gives the corresponding sulfoxide in 12% yield and benzaldehyde, the product of α C-H deprotonation, in 37% yield and the products of C-S cleavage in 26% yield. Diphenyl disulfide is also produced in 8% yield. Such photooxidation of **80b** produces no sulfoxide, only the products of α C-H deprotonation and C-S bond cleavage. It is suggested that the SOMO of the radical cation produced in this case may be localized on the $p\text{-MeOC}_6H_4$ ring and not on sulfur. This would account for the lack of sulfoxide formation and is supported by analysis of the photoelectron spectrum of **80b**. Photooxidation of **80c** sensitized by DCA does not produce any sulfoxide, even though a sulfur radical cation is produced, but only the products of C-S bond cleavage. This result shows that unimolecular C-S bond cleavage, in which a tertiary benzylic cation is formed, is faster than bimolecular reaction of the sulfur radical cation and O_2^{-}. Irradiation of sulfonium salt **81** results in intramolecular rearrangement to a mixture of sulfides **82** [197].

82

Related photochemically induced rearrangements have also been reported [198, 199]. The mechanism suggested for this reaction is photochemically induced transfer of an electron from the anthracene moiety πMO to the MeS σ^* MO resulting in C-S cleavage. This produces the caged sulfur radical cation and p-cyanobenzyl radical as a singlet radical pair. Radical-radical coupling followed by proton loss gives **82**. Photochemical electron transfer generates sulfur radical cation **83** which undergoes C-C bond cleavage resulting in the production of stabilized fragments as shown in Eq. (38) [200]:

$$\underset{\underset{SPh}{|}}{\overset{\overset{OH}{|}}{PhCHCHPh}} + A^{\bullet} \longrightarrow \underset{\underset{{}^{+}_{.}SPh}{|}}{\overset{\overset{OH}{|}}{PhCHCHPh}} + A^{-} \longrightarrow \underset{}{\overset{\overset{O}{\parallel}}{PhCH}} + \underset{\underset{SPh}{|}}{\overset{.}{CHPh}} + AH \cdot \tag{38}$$

Similar fragmentation occurs on irradiation of β-phenylthioalcohols such as **84a** with benzophenone as sensitizer, resulting in the formation of products such as **85a** in 40–93 % yield [201]. The mechanism for this reaction involves the formation of the corresponding sulfur radical cation by electron transfer from the sulfide to excited state benzophenone (a process which will be discussed in more detail in a subsequent section). The benzophenone anion radical formed in this process then deprotonates the alcohol moiety concomitantly with C=O bond formation and C-C bond cleavage as indicated in **86**.

84a, R=H   **85a**, R=H   **86**   **87**
     b, R=OH   **b**, R=OH

Support for this mechanism is obtained by the observed >20× greater reactivity of **88** over **89**. In **88** the contact radical ion pair has a geometry conducive to deprotonation but this is not the case for **89**.

$$
\begin{array}{cc}
\text{Me} & \text{OH} \\
\text{cyclohexane—OH} & \text{cyclohexane—Me} \\
\text{SPh} & \text{SPh} \\
\textbf{88} & \textbf{89}
\end{array}
$$

Consequently back electron transfer is more favorable than deprotonation for the contact radical ion pair formed with **89**. Interestingly, compound **84b** undergoes monocleavage on irradiation with benzophenone because **85b** formed after the first cleavage is protected from further reaction by forming hemiacetal **87** [202]. This rearrangement has been extended and used for the synthesis of deoxypyranoses and furanoses in 40–65 % yields. Irradiation of oxidized flavin generates the triplet excited state which is quenched by one-electron donation from $PhSCH_2CO_2H$ to form $Fl^{\cdot -}$ and $Ph\overset{+}{S}CH_2CO_2H$. Proton transfer gives FlH· and $Ph\overset{+}{S}CH_2CO_2^-$, which decarboxylates to yield $PhS\dot{C}H_2$. In addition to $PhSCH_3$, 4a-FlHCH$_2$SPh is isolated as a product in this reaction [203].

Pulse radiolysis techniques, which will be discussed in more detail in a subsequent section, have been used to study selected alkyl aryl sulfide radical cations [145]. Oxidation of the corresponding sulfides with $SO_4^{\cdot -}$ or Tl^{2+} in water gave the radical cations whose optical absorption spectra were mentioned above. These radical cations do not undergo S,S 2c, 3e bond formation with the parent sulfide. Since these radical cations are produced in very low concentration, under these conditions, dimerization, as is observed electrochemically in anhydrous acetonitrile, is disfavored. Thioanisole radical cation has a lifetime under these conditions of >30 ms but in the presence of $^-$OH the rate constant for α C-H deprotonation is 3.4×10^7 $M^{-1}s^{-1}$. The radical cations of **58c–e** rapidly undergo α C-H deprotonation and C-S bond cleavage, e. g., for the radical cation of **58e** the rate constants for both of these processes is 1.3×10^3 s^{-1}. In the presence of $^-$OH, α C-H deprotonation of the radical cation of **58d** is the predominant pathway with a rate constant of 9.5×10^6 $M^{-1}s^{-1}$, which surprisingly is less than that for thioanisole radical cation. When the radical cation of thioanisole is produced by reduction of the corresponding sulfoxide under pulse radiolysis conditions, the reactions of this species with nucleophiles such as I$^-$, N$_3^-$, PhS$^-$, PhSH, Br$^-$, and $^-$SCN, which would be oxidized with $SO_4^{\cdot -}$ or Tl^{2+}, could be studied. Nucleophilic attack was not observed but rather electron transfer. The rate of electron transfer is diffusion controlled, about 10^{10} $M^{-1}s^{-1}$, for the reactions with I$^-$, N$_3^-$, PhS$^-$, and PhSH, and slower for the reactions with $^-$SCN and Br$^-$, 2.4×10^8 $M^{-1}s^{-1}$ and 8.0×10^5 $M^{-1}s^{-1}$, respectively. No reaction was observed for NO$_3^-$ (k<10^6 $M^{-1}s^{-1}$). The lack of reaction of NO$_3^-$ is surprising because, as pointed out above, alkyl aryl sulfide radical cations, generated by Ce(IV) oxidation in AcOH, react with NO$_3^-$ to form the corresponding sulfoxides [110, 169].

The possible intermediacy of sulfur radical cations on enzymatic oxidation of phenylthioethers has attracted considerable interest. A number of enzymes are known to catalyze the oxidation of sulfides to sulfoxides [204]. Such oxidations may proceed by a one-step, two-electron oxidation concomitant with atom transfer or by initial one-electron transfer forming an intermediary sulfur radical cation. An electron-transfer mechanism has been suggested [205] for cytochrome P-450 oxidation of thioanisole derivatives ArSMe. The evidence for this suggestion is that the rate of enzymatic oxidation as a function of substituents in the Ar moiety correlate with the Hammett substituent parameter σ^+ and with the anodic peak potential for ArSMe [206]. Furthermore dealkylation of sulfur on enzymatic oxidation of ArSCH$_2$X (to produce O=CHX) increases at the expense of sulfoxide formation as X becomes more electron-withdrawing [207–209]. This change in product distribution is consistent with partitioning of Ar$\overset{+\cdot}{S}$CH$_2$X as shown in Scheme 5, paths a and b. Dealkylation at sulfur requires α-deprotonation and increasing the electron-withdrawing ability of X is expected to render Ar$\overset{+}{S}$CH$_2$X more acidic. Such increasing acidity favors dealkylation at the expense of sulfoxide formation. However, this evidence has been challenged. In view of the modest dependence of the rate of enzymatic oxidation on substituents ($\rho = -0.16$), the significance of the correlation with σ^+ has been questioned [210]. Furthermore, substituent effects do not distinguish between two-electron atom transfer oxidation and one-electron oxidation [211]. In addition, an alternative explanation has been offered [210] for the competition between dealkylation of sulfur and sulfoxide formation on enzymatic oxidation of ArSCH$_2$X. That is, dealkylation of sulfur occurs via direct hydrogen abstraction rather than by one-electron-transfer followed by α-deprotonation. Consequently an additional mechanistic probe was developed. As pointed out above, the radical cations of **80b,c** undergo C-S bond cleavage (path c in Scheme 5) in addition to α C-H deprotonation. Therefore, such unimolecular C-S bond cleavage of these radical cations is suggested to be diagnostic for the formation of such species. However, oxidation of **80b,c** with microsomal cytochrome P-450 gave exclusively the corresponding sulfoxides and sulfones and no products from α C-H or C-S bond cleavage [212]. While this test suggests that one-electron transfer does not occur with cytochrome P-450, a different result was found for oxidations with horseradish peroxidase.

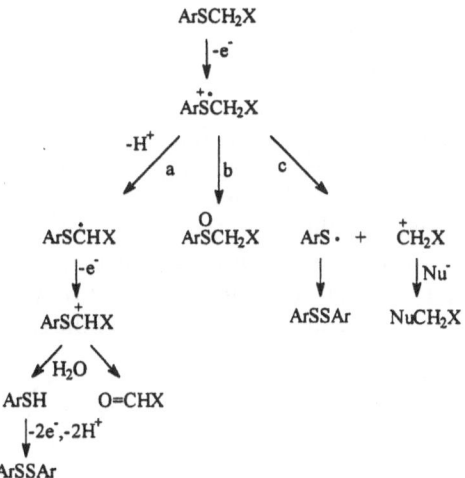

Scheme 5

Horseradish peroxidase is a heme protein that catalyzes the oxidation of a variety of substrates by one-electron transfers. It is oxidized by H$_2$O$_2$ to compound I, which is an Fe^{IV}=O porphyrin radical cation. On one-electron reduction compound I forms compound II, which is an $Fe^{IV}=O$ species. One-electron reduction of compound II yields the Fe^{III} resting state of the peroxidase. The only two-electron oxidation that horseradish peroxidase has been shown to catalyze is the oxidation of thioanisoles to the corresponding sulfoxides. It has been suggested that this oxidation occurs by an initial one-electron transfer because the rate of reaction of compound II with substituted thioanisoles correlates with σ^+ with a large substituent effect ($\varrho = -1.46$) [213], with anodic oxidation potentials (for thioanisole, $PhCH_2SMe$, and $PhCSNH_2$) [214] and compound II has been observed as a transient in this reaction [215, 216]. Subsequent to the formation of a sulfur radical cation, sulfoxide is formed either by an oxygen rebound mechanism, i.e., ferryl oxygen transfer, or coupling of $\cdot OH$, generated from compound II with the sulfur radical cation; in either case the source of the oxygen is H_2O_2 as shown by O-18 labeling studies. Ferryl oxygen transfer is unusual for horseradish peroxidase but the thioanisole binding site has been shown [217] to be different from the binding site of other substrates. This binding site is sterically restricted, but can be made more accomodating in mutants in which leucine or threonine replace phenylalanine-41 [218]. To test for the intermediacy of radical cations in oxidations of thioanisoles catalyzed by horseradish peroxidase the diagnostic C-S bond cleavage, described above for cytochrome P-450 catalyzed oxidations, was used. However, instead of using **80b,c** the more soluble **58d,e** were employed. The rates of α C-H deprotonation and C-S bond cleavage for the corresponding radical cations of these compounds, studied by pulse radiolysis techniques, is described above. Oxidation of these compounds by H_2O_2 catalyzed by horseradish peroxidase [219] gave, in addition to the corresponding sulfoxides, the α C-H deprotonation and C-S bond cleavage products in a similar ratio to that obtained by chemical oxidation supporting the intermediacy of radical cations in this enzymatic oxidation.

1.5
Thioalkene Radical Cations

Studies on the radical cations of vinyl sulfides are rather scarce. The first anodic oxidation potentials of vinyl sulfides **90–94** [220–222] are tabulated in Table 6 and are irreversible. Linear correlation of substituted-phenyl vinyl sulfides **90**, listed in the last column of Table 6, performs well with σ^+ but less well with σ. Since electrophilic additions to substituted-phenyl vinyl sulfides correlate better with σ than σ^+, it has been suggested [221] that σ^+ vs σ correlation can be used to distinguish electron transfer from electrophilic reactions.

The effect of C-Si and C-Sn bonds, adjacent to the sulfur of vinyl sulfides, on the oxidation potential has been evaluated [222]. There is only a modest effect for a C-Si bond but (E)-**94**, $R=(c\text{-}C_6H_{11})_3Sn$, $R^1=t\text{-}Bu$, $R^2=H$ has an oxidation potential

Table 6. Oxidation potentials[a] of vinyl sulfides in acetonitrile

Compound	$E_p{}^b$, V vs Ag/0.01 M Ag$^+$, CH$_3$CN	$E_p{}^c$, V vs Ag/0.1 M Ag$^+$, CH$_3$CN	$E_{ox}{}^d$, V vs SCE
90, R$_m$=R$_p$=H	–	1.36	1.346
90, R$_m$=H,R$_p$=MeO	–	–	1.097
90, R$_m$=H,R$_p$=Me	–	–	1.267
90, R$_m$=H, R$_p$=Ph	–	–	1.249
90, R$_p$=H, R$_m$=Me	–	–	1.314
90, R$_m$=H, R$_p$=Cl	–	–	1.401
90, R$_m$=H, R$_p$=Br	–	–	1.389
90, R$_p$=H, R$_m$=Cl	–	–	1.475
90, R$_m$=H R$_p$=CF$_3$	–	–	1.594
91	–	–	1.603
92, R=R^1=Ph	0.97	0.89^e, 0.96^f	–
92, R=Ph, R^1=t-Bu	–	1.01^e, 1.04^f	–
92, R=t-Bu, R^1=H	–	1.27	–
92, R=n-Pr, R^1=Ph	0.87	–	–
92, R=n-Bu, R^1=Ph	0.88	–	–
92, R=i-Bu, R^1=Ph	0.88	–	–
92, R=s-Bu, R^1=Ph	0.88	–	–
93	–	1.03	–
94, R=PhCH$_2$,R^1=H,R^2=Ph	1.11	–	–
94, R=H, R^1=Ph, R^2=CN	1.38^e, 1.31^f	–	–
94, R=Ph, R^1=H, R^2=CN	1.60	–	–

[a] First oxidation potential   [c] [222]   [e] E isomer
[b] [220]   [d] [221]   [f] Z isomer.

560 mV lower than that of phenyl vinyl sulfide. Since (E)-**94**, R=(c-C$_6$H$_{11}$)$_3$Sn, R^1=t-Bu, R^2=H has a lower oxidation potential than **94**, R=(c-C$_6$H$_{11}$)$_3$Sn, R^1=R^2=H, the orientation of the C-Sn bond with respect to the p-orbital on sulfur is suggested [222] to be important. X-ray crystallographic structure studies show that **94**, R=(c-C$_6$H$_{11}$)$_3$Sn, R^1=R^2=H adopts a *cis* conformation **95** in which the C-Sn bond and sulfur p-orbital are orthogonal; whereas, the *t*-Bu group of (E)-**94**, R=(c-C$_6$H$_{11}$)$_3$Sn, R^1=t-Bu, R^2=H precludes the *cis* conformation and a gauche conformation **96** is adopted in which there is substantial overlap between the C-Sn bond and sulfur p-orbital (the C(Ph)SC(1)=C(2) torsion angle is 139° in the solid state).

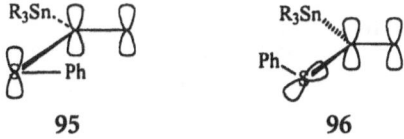

         95   **96**

Controlled potential electrolysis of **92**, R=R^1=Ph; **92**, R=n-Pr, R^1=Ph; **92**, R=n-Bu, R^1=Ph; **92**, R=i-Bu, R^1=Ph; and **92**, R=s-Bu, R^1=Ph surprisingly results in apparent deprotonation β to the sulfur in an ECE process to give **97**, R=Ph; **97**, R=nPr; **97**, R=n-Bu; **97**, R=i-Bu; and **97**, R=s-Bu, respectively, in about 50% yield via the corresponding thiirenium salts [220].

PhCHCHO   PhC(CN)CH(SPh)CN   PhC(CN)=C(SPh)$_2$
| |
SR   MeO
97   **98**   **99**

When such deprotonation is precluded, as in **94**, R^1=Ph, R^2=CN, nucleophilic addition of added methanol to the radical cation activated by the electron withdrawing CN group occurs to give **98** as the main product isolated in 43% yield. Controlled potential electrolysis of **94**, R=H, R^1=Ph, R^2=CN produces **99** in modest yield (34%). It is suggested that this product results via dimerization of the intermediary radical cation.

Chemical oxidation of (E)-**94**, R=(c-C$_6$H$_{11}$)$_3$Sn, R^1=t-Bu, R^2=H using the one-electron oxidants Mn(III) acetate and ammonium cerium(IV) nitrate resulted in C-Sn bond cleavage in the radical cation [222]. The products in these reactions are PhSC≡Ct-Bu, (Z)-**92**, R=Ph, R^1=t-Bu and/or (E)-**92**, R=Ph, R^1=t-Bu.

The radical cations of polythioalkenes are well known. The EPR spectra of the radical cations of dithioalkenes **100–104** [94], tetrathioalkenes **105–106** [94, 223–226], dibenzotetrathiafulvalene **107** [226], and tetrathiafulvalene (TTF) **43** [225–228] generated in sulfuric acid, by AlCl$_3$ in CH$_2$Cl$_2$, anodic oxidation, NOBF$_4$ iodine or chlorine oxidation show g$_{av}$ values from 2.0077–2.0193.

100a, R = R^1=H   **101a**, R = H   **102**
   b, R = Me, R^1=H   **b**, R = Me
   c, R = R^1=Me

103   **104**

105a, R = Me   **106a**, n = 0   **107**
   b, R = Et   **b**, n = 1
   c, R = i-Pr
   d, R = Ph

Many of the polythioalkenes have been studied electrochemically. Compounds **43**, **106b**, and **108** show two reversible one-electron oxidations as determined by cyclic voltammetry in acetonitrile [225]. Compounds **105d** and **106a** show a reversible oxidation for the first step but the second step is irreversible suggesting that the dications in these cases react with the solvent. The dications

Table 7. Oxidation potentials for tetrathioalkenes in acetonitrile

Compound	$E_{1/2}$, V^a vs SCE	$E^{0'}$, V^b vs SCE	$E_{1/2}$, V^c vs SCE
105a	0.89	–	–
105b	0.93	–	–
105c	0.97	–	–
105d	1.18	1.2	–
106a	–	0.68, 1.12	0.90, 1.03
106b	–	0.695, 0.84	0.72, 0.90
107	–	0.33, 0.70	–
108	–	0.405, 0.89	–
109	–	–	0.90, 1.28
110	–	–	0.71, 1.17

[a] [223]   [b] [225]   [c] [229, 230].

of **109** and **110** undergo the rearrangement shown in Eq. (39) [229, 230]. It is suggested that both σ bonds migrate simultaneously in this rearrangement whose driving force is the lowering of electrostatic repulsion between the positively charged sulfur atoms. In **111** rotation about the central C-C bond, which is not possible in dications of **109** and **110**, minimizes this repulsion. This rearrangement is analogous to that outlined above for the dication of **40a** giving the dication of **43**. The electrochemical data are recorded in Table 7.

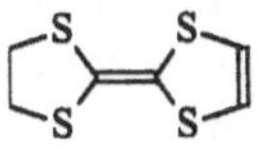

108 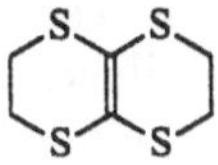   109 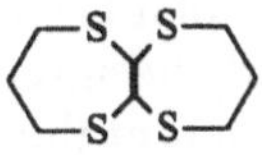   110

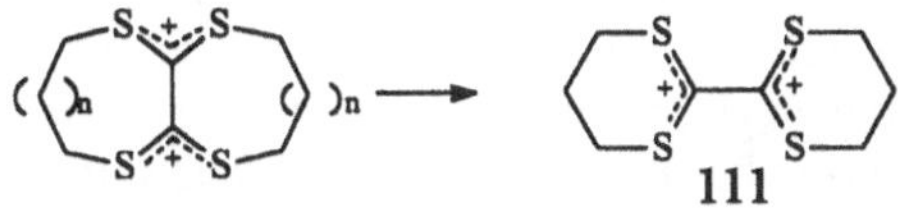

(39)

111

Charge transfer complexes of TTF, **43**, and its derivatives with tetracyanoquinodimethane (TCNQ) and other π-acids are highly electrically conductive [231], justifying their designation as organic metals, when they crystallize in separate stacks. Partially oxidized TTF and its derivatives prepared by chemical oxidation or electrocrystallization techniques also show metal-like conductivity when they crystallize with molecules arranged in π-stacks. At low temperatures some of these species become superconducting. For example, a number of salts of **112** show superconducting transition temperatures between 1 to 13 K [232, 233].

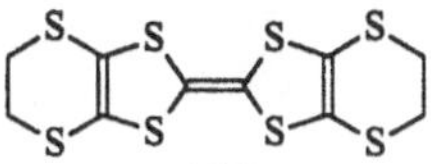

112

Consequently, there has been an enormous interest in preparing TTF derivatives, studying their redox behavior, determining the structure and metal-like conductivity of their charge transfer complexes and partially oxidized salts. This

work has been extensively reviewed [234–237] and the interested reader is referred to these reviews for insight into this exciting area. The challenge to our
understanding in this area is to provide the link between molecular properties
and the colligative properties of materials. That is, unify the chemistry and phy-
sics of organic metals and superconductors. Since this review focuses on the
molecular properties of sulfur radical cations the reader interested in the link-
age between molecular and material properties is referred to insightful essays
on this subject [238, 239].

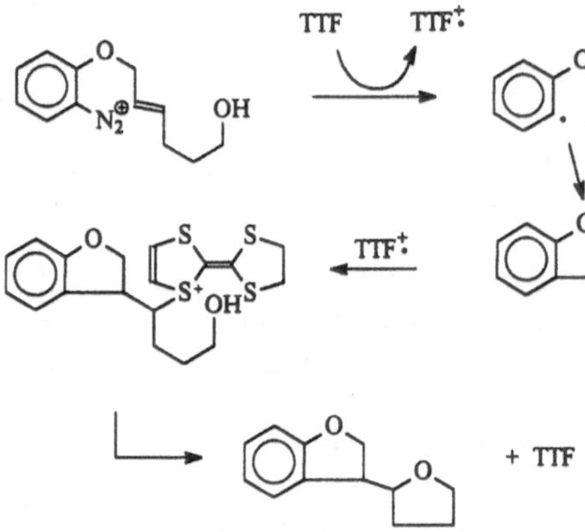

Scheme 6

The reaction of TTF$^{\ddot{+}}$ with radicals resulting in the formation of sulfonium
salts has recently been reported. An illustrative example is shown in Scheme 6
[240] and a review of the synthetic methodology has appeared [241]. The first
step involves electron transfer from TTF to the diazonium salt resulting in the
formation of TTF$^{\ddot{+}}$ and the aryl radical shown which cyclizes. The resulting cyclized radical couples with TTF$^{\ddot{+}}$ to give the sulfonium salt shown. An alternative
mechanism involving aryl cations instead of aryl radicals and nucleophilic
attack by TTF on a cation has been refuted [242]. The sulfonium salt then undergoes nucleophilic displacement by an S_N1 mechanism. Thus radical reactions are
coupled with ionic reactions in a synthetically advantageous way.

1.6
Thiophene Radical Cations

The radical cations of thiophene **113a**, and its derivatives have been made in
Freon by γ-radiolysis [243, 244] and in solution by UV irradiation [245].

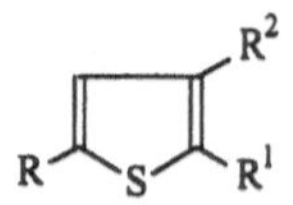

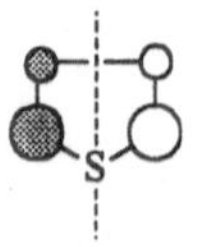

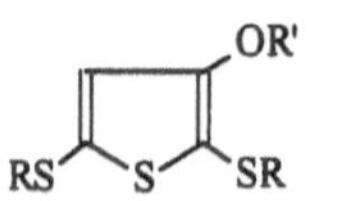

113a, R = R^1 = R^2 = H   **114**   115a, R = (CH$_2$)$_2$OAc R′ = Me
  b, R = Me, R^1 = R^2 = H   b, R = Me, R′ = (CH$_2$)$_4$CH$_3$
  c, R^2 = Me, R = R^1 = H
  d, R = R^1 = Me, R^2 = H

The reported g_{av} values are 2.0019–2.003 [244, 245] and the hyperfine splitting constants a_H are 13 and 11.8 G [243] for H (2,5) and 3.1–3.75 G [244, 245] for H (3,4). These parameters are very similar to those for furan and pyrrole radical cation leading to the SOMO assignment as $^1a_2$. This orbital, shown in 114, has a node through the sulfur atom rendering it distinct from the other sulfur radical cations considered in this review because there is no spin density on sulfur. Thiophene radical cations are very unstable in solution and readily polymerize. This polymerization is suggested [246] to occur via coupling of thiophene radical cations as shown in Scheme 7. Rearomatization via proton loss generates 2,2'-bithienyl which is more easily oxidized than thiophene. The resulting radical

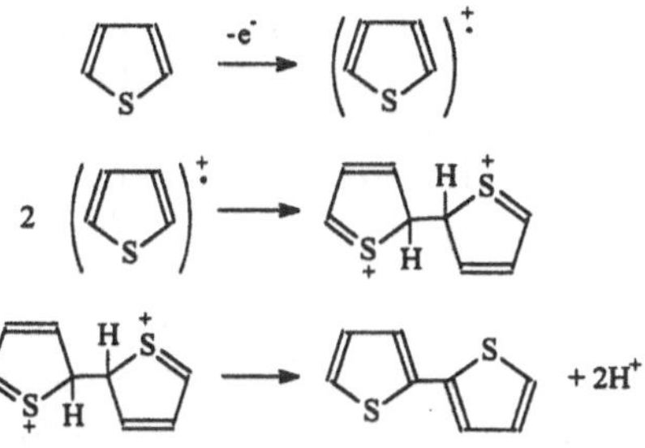

Scheme 7

cation can react with thiophene radical cation and a series of $E(CE)_n$ processes generates polythiophene. Polythiophene is of interest as a stable electrical conductor and semiconductor especially when doped. This material may be of value for "plastic electronics" in which lightweight, processable, deformable plastics replace metals [247, 248]. Since the SOMO coefficient for thiophene radical cations is greater at the α-position than β-position, α,α'-coupling occurs as shown. This is in accord with the experimental spin density distribution obtained from the EPR spectrum using the McConnell equation and density functional theory calculations [249]. Consequently, α-substituted thiophene radical cations are more stable than the parent species especially with α-alkylthio groups. Thus, 3-alkoxy-2,5-bis(alkylthio) thiophene 115a,b showed reversible one-electron oxidations at 0.79 and 0.53 V vs SCE, respectively in 1,1,1,3,3,3-hexafluoropropan-2-ol [250]. In addition the EPR spectra for 115a,b were measured showing a_H=4 G (1H), a_H=ca. 2 G(7H) and a_H=4 G(1H), a_H=ca. 2 G(8H) respectively.

The basis for electrical conductivity in doped polythiophene is of interest. Either polarons (delocalized radical cations) or dipolarons (delocalized dications) have been suggested. To ascertain the properties of these species oligothiophenes have been prepared and studied. These species not only serve as models for polythiophenes, but are also of interest themselves as field effect transistors and light emitting diodes [247, 251]. Indeed the optimum field effect mobility is attained with six α-linked thiophenes, α-sexithienyl, even surpassing the homologous polymer [252]. The oligothiophenes can be prepared chemically pure and defect-free unlike the polymer. Furthermore, α-terthienyl and related compounds are secondary plant metabolites that are phototoxic and may be of value as insecticides [253]. Therefore a brief overview of oligothiophene radical cations is presented here.

116a, R=R¹=R²=H
  b, R=Me, R¹=R²=H
  c, R=i-Pr, R¹=R²=H
  d, R=t-Bu, R¹=R²=H
  e, R=Me₃Si, R¹=R²=H
  f, R=Me, R¹=H, R²=OMe
  g, R=Me, R¹=OMe, R²=H

117a, n=1, R=R'=R''=R'''=H
  b, n=1, R=Me, R'=R''=R'''=H
  c, n=1, R=Me, R'=R'''=H, R''=OMe
  d, n=1, R=Me, R'=OMe, R''=R'''=H
  e, n=1, R=Ph, R'=R''=H, R'''=Bu
  f, n=2, R=Me, R'=R'''=H, R''=OMe
  g, n=2, R=Me, R''=R'''=H, R'=OMe
  h, n=3, R=Me, R'=R''=H, R''=OMe

2,2'-Bithienyl **116a** exists in two conformations in the gas phase as shown by electron diffraction studies [254].

The ratio of these *trans* to *cis* isomers is 56:44 and their torsion angles are 148 and 36°, respectively. Theoretical and experimental studies suggest that 4,4'-dialkyl-2,2'-bithienyl derivatives are similar to the unsubstituted compound but 3,3'-dialkyl-2,2'-bithienyl derivatives are more twisted [255]. The radical cations of 5,5'-disubstituted-2,2'-bithienyl **116b–e** have been generated by photolysis in CF_3CO_2H with or without Hg(II) or Tl(III) trifluoroacetate [256] and their EPR spectra measured. Two radical species are formed in the ratio of ca. 3:2, which are suggested to be the *trans* and *cis* conformations. The barrier for interconversion of these conformers is greater than that in the parent compounds. The g-values are in the range of 2.00285–2.0062 and these low values are consistent with little spin density on the sulfur atom. The hyperfine splitting constants a_H are 3.77–3.91 for H(3,3') and 0.70–1.08G for H(4,4'). The irreversible first oxidation potentials for **116b–e** in DMF determined by cyclic voltammetry are 1.21, 1.26, 1.27, and 1.44 V vs SCE [256] and the value for 2,2'-dithienyl itself **116a** under these conditions is 1.41 V and in CH_3CN it is 1.32 V and **116e** is 1.14 V [257]. Without 5,5'-substituents 2,2'-bithienyl radical cations dimerize under these conditions. However, 2,2'-bithienyl radical cation can be generated photochemically [253]. Irradiation of 2,2'-bithienyl in CH_3CN generates the triplet excited state which is quenched by methyl viologen or TCNE by electron-transfer at near diffusion-controlled rates. This yields the corresponding radical cation which absorbs with λ_{max} 420 and 580 nm. This radical cation is relatively unreactive toward O_2 but triplet **116a**, like triplet α-terthienyl, is an excellent singlet oxygen sensitizer. The radical cation of **116a** can also be formed by oxidation with ZSM-5 or Na-β zeolites [258]. 2,2'-Bithienyl derivative **116f** undergoes reversible one-electron oxidation in CH_2Cl_2 to give a stable radical cation but its isomer **116g** does not afford a stable radical cation [259]. This difference between the two isomers is consistent with the corresponding MO coefficients at the carbon bearing the stabilizing MeO group. α-Terthienyl radical cation can be formed photochemically [253–260]. Triplet **117a** is quenched by electron-transfer with methyl viologen or TCNE in an analogous way as discussed for **116a**. The radical cation of **117a** absorbs at 530 nm. Zeolites also oxidize **117a** to produce the corresponding radical cation identified by its diffuse reflectance and EPR spectra [258]. Anodic oxidation of α-terthienyl **117a** is irreversible because its radical cation couples forming α-octathienyl (and two protons) which precipitates on the electrode [261]. However, blocking the reactive α-positions increases the stability of the radical cation. Thus anodic oxidation of **117b** in CH_3CN gave the corresponding radical cation characterized by EPR spectroscopy [262]. Unambiguous assignment of the β-hyperfine splitting constants were made for this radical cation using deuterated compounds. Terthienyl **117c** shows two reversible one-electron oxidations in CH_2Cl_2 as shown by cyclic voltammetry with E^0 values of 0.50 and 0.92 V vs SCE [259]. The corresponding radical cation is stable for more than one month in CH_2Cl_2 solution or as an isolated salt. However, its isomer **117d** does not form a stable radical cation. This result is similar to that reported for 2,2′-bithienyl derivatives **116f,g**. α-Terthienyl derivative **117e** shows two reversible one-electron oxidations in CH_2Cl_2 at 0.89 and 1.29 V vs Ag/AgCl by cyclic voltammetry [263]. Analysis of the absorption spectrum of the radical cation of **117e** showed π-dimerization with K_{eq} at 295 K about 10^{-4} in CH_2Cl_2 and 0.6 in CH_3CN. Similar dimerization was reported [264] for the radical cation of **117b** with K_{eq} at room temperature of ca. 154 in CH_3CN. Dimerization is suggested to be favored in CH_3CN relative to CH_2Cl_2 because the more polar CH_3CN can better reduce the Coulombic repulsion in the dication [259]. EPR spectroscopic studies [263] on **117e** radical cation show two species, suggested to be rotational isomers, with g-values of 2.0014–2.0025 depending on temperature. Futhermore, on cooling the intensity of the EPR signal decreases due to reversible π-dimerization of the radical cation to give a diamagnetic dication. The X-ray crystal structure of the radical cation of **117e** shows π-stacks evenly spaced by 3.47 Å at 293 K but at 106 K the π-stacks consist of π-dimers with regularly alternating distances of 3.36 and 3.42 Å. This π-stacking may be of importance for electron conduction both in this species as well as in other oxidized oligo- and polythiophenes. This is an alternative mechanism for electron conduction in addition to the polaron/bipolaron mechanism. This alternative mechanism provides for electron transfer between chains, thereby increasing the dimensionality of the conduction. Other mechanisms for achieving this have been suggested as well. Transient formation of polarons from bipolarons has been suggested to be energetically feasible and provides a way for interchain charge hopping [258]. The dimerization of radical cations is of significance in determining whether polarons or bipolarons are the principal species in oxidized polythiophene. The low number of spins in oxidized polythiophene has been taken as evidence for the dominance of bipolarons. However, radical cation π-dimers are diamagnetic and are also suggested [263] to account for the low number of spins. Similarly cyclic voltammetric studies in CH_2Cl_2 on soluble α-terthienyl derivative **118a**, end substituted to stabilize the oxidation products, undergoes reversible one-electron oxidation to the corresponding radical cation with E^0 of 0.38 V vs Fc/Fc^+ and irreversible oxidation at 0.79 V to the dication [265]. Cyclic voltammetric studies on **119a,b** and **120a** in CH_2Cl_2 each show two oxidation potentials with E_p of 1.21, 1.57; 1.16, 1.58; and 1.17, 1.53 respectively [266]. The first oxidation potential for **120b** is reversible but it is not for **120a** which apparently dimerizes on oxidation. No spectroscopic evidence is found for the formation of π-dimers.

118a, n = 1
  b, n = 2
  c, n = 3
  d, n = 4

119a, n = 0, R = H
   b, n = 0, R = Me
   c, n = 2, R = H
   d, n = 2, R = Me

120a, n = 0
   b, n = 2

Two electrochemically reversible oxidation steps are found by cyclic voltamme-try in CH_2Cl_2 for **117f,h** with E^0 of 0.49, 0.77 V and 0.48, 0.70 V vs SCE, respec-tively. Radical cation **117f** is completely stable in CH_2Cl_2 solution and for more than one month as its isolated salt. Both **117f,h** radical cations undergo π-dimer-ization with K 79 mM^{-1} and $>10^3$ mM^{-1}, respectively. Cyclic voltammetric studies on **117g** showed irreversible oxidation with a peak potential about 0.7 V in contrast to its isomer **117f**. Compounds **118b–d** show two reversible one-elec-tron oxidations at E_1^0 of 0.32, 0.26, and 0.22 and E_2^0 of 0.66, 0.55, and 0.41 V vs Fc/Fc$^+$, respectively [265]. The radical cations of **118b–d** undergo reversible π-dimerization on cooling to 200–235 K as determined spectroscopically. These dimers are spin paired as shown by EPR spectroscopy, i.e., the number of spins decreases on cooling. This dimerization is more favorable as the π-conjugation increases: ΔH = –42, –58±5, –65±3, and –87±16 kJ/mol for dimerization of the radical cations of **118a–d**, respectively. This dependence of dimerization on increasing π-conjugation, in these examples as well as those for the radical ca-tions of **117c,f,h**, is ascribed to decreasing Coulombic repulsions in the dication. However, studies on the radical cations of **119a–d** and **120a,b** show no evidence for π-dimers [266]. Spiro compound **120b** can be selectively oxidized with $FeCl_3$ to the monoradical cation, bis-radical cation, radical cation/dication, and bis-dication. The spectroscopic correspondence between the oxidized forms of **119d** and **120b** show that there is no significant interaction between the orthogonal oxidized oligothiophenes. However there appears to be interaction between the radical cation and uncharged oligomer moieties because the first and second oxidation potentials differ for **119d** and **120b**. The electrochemistry of variously substituted α-sexithienyls has been reported [267] as well as photochemical oxi-dation of α-sexithienyl [260].

121a, n=1   122a, n=1, x=1
  b, n=3   b, n=2, x=0

Substituted α-quaterthienyl **121a**, α-octithienyl **122a**, α-dodecithienyl **121b**, and α-sedecithienyl **122b** have been prepared and their peak potentials determined by cyclic voltammetry in CH_2Cl_2 are 0.51, 0.34, 0.19, and 0.12 V vs Fc/Fc^+ [268]. The oxidation potentials for **121b** and **122b** are less positive than those for poly-thiophene and polyalkylthiophene (0.30–0.35 V) suggesting greater delocaliza-tion in these compounds than in the polymers (where the average chain length for π-conjugation is estimated as 8–10 monomer units). Photochemical oxida-tion of α-quaterthienyl in CH_2Cl_2 has also been reported [260]. The radical ca-tions of these oligothiophenes show increasing tendency to dimerize with increasing chain length as noted above. In spiro compounds **119d** and **120b** with orthogonal oligothiophene chains, there appears to be interaction between the radical cation and neutral oligomer moieties as pointed out above. Electro-chemical evidence for interaction between the radical cation of one oligothio-phene chain and a nearby cofacially oriented neutral oligothiophene chain has been presented [269]. The peak potentials for irreversible oxidation of **123a–f** determined by cyclic voltammetry in CH_3CN are 1.11, 1.05, 0.84, 0.75, 0.71, and 0.59 V vs Ag/Ag^+, respectively.

123a, m=1, n=0   d, m=n=2
  b, m=n=1   e, m=3, n=0
  c, m=2, n=0   f, m=n=3

Pairwise comparison of **123a** with **b**, **c** with **d**, and **e** with **f** show that the oxida-tion potential is rendered less positive by 60, 90, and 120 mV, respectively, when there is a matching oligothiophene chain to its absence. Connection of oligothiophene moieties by ethylene units as in **124** [270] results in greater interaction than connection by disilanylene units as in **125** [271]. Another ex-ample of thiophenes connected by an ethylene unit is tetrathienylethene **126**. Cyclic voltammetric studies in CH_3CN show irreversible oxidation with peak potentials of 0.91, 0.70, and 0.55 V vs SCE for **126a–c**, respectively [272]. Electro-lysis of **126c** in CH_2Cl_2 gives the corresponding radical cation whose EPR spec-trum was measured.

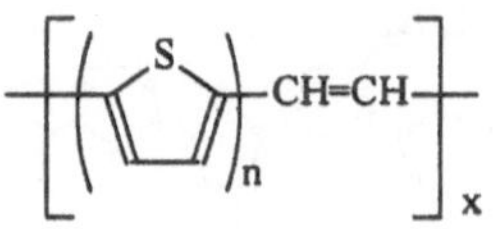

124 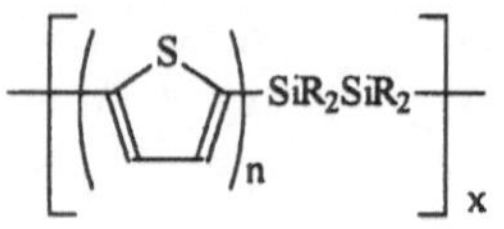   **125**

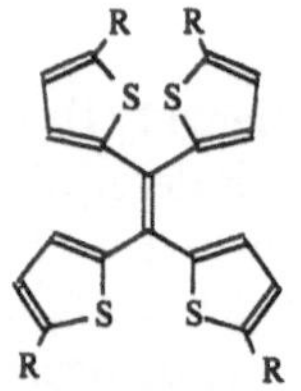

126a, R = H   **127**   **128**
b, R = Me
c, R = SMe

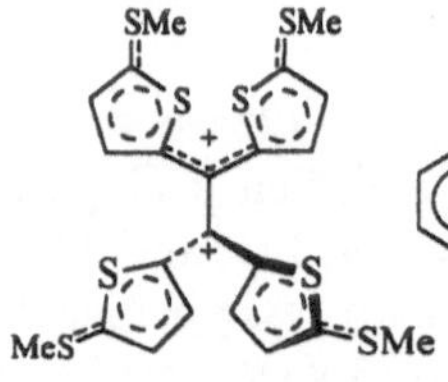

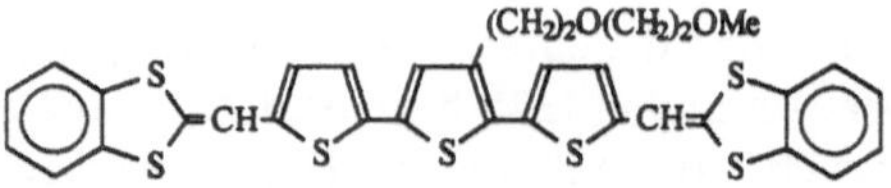

The hyperfine splitting constants a_H are 2.24 (4H, H3), 0.16 (4H, H4), and 0.62 G (12H, Me) showing that the spin is delocalized over four equivalent thiophene rings. Anodic oxidation also produces the dication of **126c** isolated as its perchlorate salt. The X-ray crystallographic structure shows that there are two delocalized dithienyl methylene units perpendicular to each other as shown in **127**. Cyclic voltammetric studies of **128** in CH_2Cl_2 have been reported [273]. This compound can be viewed as a dibenzo TTF derivative connected by an α-terthienyl moiety. It shows three quasi-reversible oxidations in CH_2Cl_2 with peak potentials of 0.54, 0.70, and 0.82 V vs SCE. These may be compared with the first oxidation potentials of dibenzo TTF and 3′-(3,6-dioxaheptyl)ter-thienyl of 0.72 V and 1.05 V, respectively.

The radical cation of dibenzothiophene **129** has been prepared in 1,1,1,3,3,3-hexafluoro-2-propanol solution by oxidation with Tl(OAc)$_3$ and its EPR spectrum measured [274].

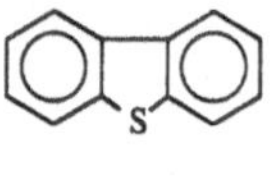

129 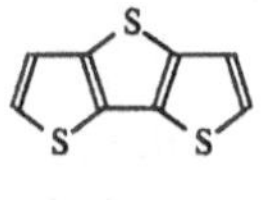   **130** 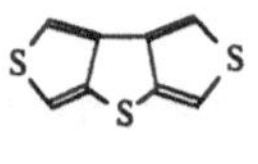   **131** 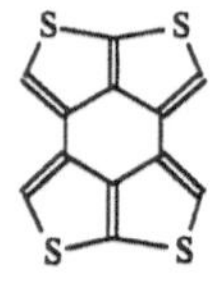   **132**

Unlike thiophene radical cation the SOMO for this species is $^2b_1$ with considerable spin density on sulfur. The reversible electrochemical oxidation potentials for **129** and some of its derivatives in 1,1,1,3,3,3-hexafluoro-2-propanol are listed in Table 8. The reactions of **129**$^{+}$ with radicals and with nucleophiles has been studied [275]. The position of attack by radicals on **129**$^{+}$ should reflect the spin density at that position as found by EPR spectroscopic analysis. Indeed reaction with NO$_2$ occurs predominantly at S, C(2) and C(4) as expected. The valence bond configuration mixing model leads to the prediction that nucleophiles should preferentially attack **129**$^{+}$ at C(1) and C(3) with little attack at S, C(2) and C(4). This is partly but not completely validated experimentally. Radi-

Table 8. Oxidation potentials for dibenzothiophenes [274]

Substituents	E, V vs Ag/AgCl
129	1.26
2,8-Me$_2$	1.12
3,7-Me$_2$	1.21
4,6-Me$_2$	1.20
2,4,6,8-Me$_4$	1.04
2,2'-bis	1.08
4,4'-bis	1.25

cal cations of thiophenes annulated to thiophenes have also been generated and studied [276]. γ-Radiolysis of **130** and **131** in Freon at 77 K produced the corresponding radical cations with λ_{max} 351, 411, and 595 nm and λ_{max} 410, 880, and 918 nm, respectively. These species were also formed in acetonitrile by laser flash photolysis in the presence of an electron acceptor such as DNB or CCl$_4$. Both singlet and triplet states of **130** and **131** undergo electron transfer at diffusion controlled rates ca. 10^{10} M^{-1}s^{-1} with DNB. The onset oxidation potentials in acetonitrile for **130** and **131** are 1.07 and 1.04 V vs SCE, respectively. Cyclic voltammetric studies on **132** in PhCN show one reversible oxidation with $E_{1/2}$ of 1.01 V vs Ag/AgCl which is comparable to that of perylene [277].

133a, R = R' = i-PrS
b, R = R' = t-BuS
c, R = t-BuS, R' = EtS
d, R = t-BuS, R' = MeS
e, R = R' = 2-thienyl

Nonclassical thiophenes **133** have been studied electrochemically and show reversible oxidation waves as reported in Table 9 [277–279]. For comparison purposes the oxidation potentials of TTF under the conditions used for **133a–d** are +0.34 and +0.71 V vs Ag/AgCl [277].

Table 9. Oxidation potentials for nonclassical thiophenes 133

Compound	E$_1$ V vs Ag/AgCl[a]	E$_2$	E$_1$ V vs SCE[b]	E$_2$	E V vs Ag/Ag^{+}[c]
133a	+0.14	+0.49	+0.19	+0.54	–
b	+0.34	+0.54	–	–	–
c	+0.16	+0.46	–	–	–
d	+0.15	+0.48	–	–	–
e	–	–	–	–	+0.61

[a] [277]   [b] In CH$_3$CN [278]   [c] In CH$_2$Cl$_2$ [279].

2
Localized Radical Cations

2.1
Dialkyl Sulfide Radical Cations

2.1.1
Generation

Dialkyl sulfide radical cations have been made from the parent sulfides by a variety of methods including γ-radiolysis in matrices at low temperature [280], pulse radiolysis [281–285], photoinduced electron-transfer [286], mass spectroscopic methods [287, 288], electrochemical oxidation, and by oxidation with one-electron chemical oxidants. A brief overview of these methods will be presented here even though most of these methods have already been mentioned previously in this review.

Irradiation of dilute solutions of dialkyl sulfides in Freon matrices, such as $CFCl_3$, at liquid N_2 temperatures with ^{60}Co γ-rays provides an excellent way to generate the corresponding radical cations under conditions in which they are relatively stable. Irradiation of $CFCl_3$ with γ-rays results in ejection of an electron leaving $(CFCl_3)^+$. The electron is rapidly scavenged by another $CFCl_3$ molecule. The $(CFCl_3)^+$ constitutes a solvent "hole" which is effectively transported through the matrix via electron transfer from another $CFCl_3$ molecule until it encounters a dialkyl sulfide molecule. The ionization potential of $CFCl_3$ is around 11.9 eV whereas that for dialkyl sulfides is 8–9 eV. Consequently, the dialkyl sulfide transfers an electron to $(CFCl_3)^+$ ending the "hole" migration. The structure of the resulting sulfur radical cation can be ascertained by EPR spectroscopic analysis. Annealing the sample can also provide information on reactions of the sulfur radical cation.

The technique of pulse radiolysis is an excellent general method for generating sulfur radical cations in aqueous solution and monitoring their reactions principally by spectroscopy and conductivity in a time resolved manner (typically in the ns-ms time frame) [281–285]. In this method a high energy electron beam in the 1–30 MeV range from an accelerator (van de Graaf generator, Linac, or Febetron) is allowed to impinge on dilute aqueous solution (typically 10^{-3} to 10^{-4} mol l^{-1}) of the sulfur compound saturated with N_2O. The high energy beam electron beam loses energy rapidly (less than 10^{-14} s) in a series of steps, each event of discontinuous energy deposition is called a spur, resulting principally in the ionization and excitation of water. This ionization and subsequent steps are shown in Eqs. (40)–(43):

$$H_2O \xrightarrow[\text{beam}]{\text{electron}} H_2O^{+\cdot} + e^- \tag{40}$$

$$e^- + nH_2O \longrightarrow e^-(H_2O)_n \tag{41}$$

$$H_2O^{+\cdot} + H_2O \longrightarrow H_3O^+ + \cdot OH \tag{42}$$

$$e^-(H_2O)_n + N_2O \longrightarrow \cdot OH + N_2 + {}^-OH \tag{43}$$

Ionization (Eq. 41) results in the formation of the radical cation of water and a low energy electron. Within 10^{-12} s this electron is solvated and H_2O^+ is deprotonated to form $\cdot OH$, which is an excellent one-electron oxidant. The solvated electron then rapidly and efficiently reacts with N_2O as shown in Eq. (43) with k=8.8×10^9 M^{-1}s^{-1} for saturated 2.4×10^{-2} mol l^{-1} N_2O aqueous solutions. This efficient scavenging removes the reducing solvated electron replacing them with oxidizing $\cdot OH$ radicals. Overall the one-electron oxidizing agent, $\cdot OH$, is produced selectively and rapidly throughout the solution in concentrations sufficiently high that subsequent reactions with sulfides produces sufficient radical cation to enable easy spectroscopic and conductimetric monitoring. The yields in this process of forming sulfur radical cations are expressed as G values, which are the number of species formed per 100 eV of absorbed energy. Instead of oxidation with $\cdot OH$, other oxidizing species can be produced under pulse radiolysis conditions such as Br_2^{-}, Cl_2^{-}, I_2^{-}, $(SCN)_2^{-}$, $N_3\cdot$, SO_4^{-}, $Ag(OH)^+$, Ag^{2+}, and Tl^{2+}. These species differ from $\cdot OH$ in oxidation potential and, in most cases, charge. Use of these various oxidizing species can provide different reaction pathways, a way of distinguishing radical cations from adducts and a method for determining the oxidation potentials of sulfur compounds in aqueous solutions as already reported for diphenyl sulfide, dimethyl disulfide, and α-lipoic acid. Such use of the reaction of Br_2^{-} with dimethyl sulfide provided the oxidation potential for the oxidation of two dimethyl sulfide molecules to 3c, 2e $(Me_2S)_2^{+}$, whose bonding was discussed above and whose chemistry is discussed below, which is E^0=1.40±0.02 V vs SHE [289] was found. Furthermore the electron-transfer equilibrium of Ph_2S^{+} and Me_2S was used to determine the potential for oxidation of Me_2S to Me_2S^{+}, for which E^0=1.66±0.03 V vs SHE [289] was found. The one-electron oxidation of dialkyl sulfides by $\cdot OH$ is not a simple electron-transfer but rather a complicated process. The first step in this process is diffusion-controlled addition to generate $R_2\dot{S}$-OH [290–292]. Typically these adducts have very short lifetimes ($\tau_{1/2}$ less than 1 μs) but, if stabilized by intramolecular hydrogen bonding as in adducts **134** and **135** [293], and perhaps the adduct of $\cdot OH$ and $S(CH_2CH_2CO_2Me)$ [294] they are more persistent.

134   **135**

Adducts **134** and **135** absorb with λ_{max} at 330 and 340 nm respectively [293], and adduct $HO\dot{S}(CH_2CH_2CO_2Me)_2$ at λ_{max}=345 nm [294]. In the gas phase no experimental or theoretical evidence [295] was found for $Me_2\dot{S}OH$ as an intermediate. However, other experimental [296] and a high level ab initio calculation [297] find a stable complex weakly bound by 13±3 and 6.0 kcal/m, respectively. Adducts **134** and **135** then form radical cations by three pathways. Ionization of ^-OH from the adduct can occur by a spontaneous or acid catalyzed process. In addition, parent sulfide can displace ^-OH leading to a 2c, 3e species $(R_2S)_2^{+}$ The adduct of $\cdot OH$ and Me_2S is reported to absorb with λ_{max} at 340 [298] or 358 nm

[289]. Solvent kinetic isotope studies at pH 6.7 suggest that $Me_2\dot{S}OH$ forms $(Me_2S)_2^{+}$ in two ways. In the first way a unimolecular process not involving proton transfer from water produces Me_2S^{+} which rapidly associates with Me_2S. In the second way, which depends on the concentration of Me_2S, Me_2S displaces ^-OH from $Me_2\dot{S}$-OH in a process involving proton transfer from the solvent water to the departing ^-OH [289, 298]. Aqueous solvation of $Me_2\dot{S}$-OH is proposed to have a profound effect on its properties accounting for the reactions outlined here which differ from those in gas phase [289]. In the gas phase, $Me_2\dot{S}$-OH is calculated to have a nonpolar S-O bond. However, in aqueous solution this bond is polarized so that heterolysis, not homolysis, is favored.

As illustrated by a number of examples in this review, photoinduced electron transfer is an effective way to form sulfur cation radicals. The excited states of a number of aromatic sensitizers such as DMN and DCA as already mentioned are good oxidants. In addition, FMN [299], 2,4,6-triarylpyrylium salts [300–302], 1-cyanonaphthalene [303], 2,3-dichloro-5,6-dicyanoquinone [304], methacridinium perchlorate [302], and methylene green [305, 306] have also been used. Titanium dioxide sensitized oxidation, which involves electron transfer from a sulfide to a photochemically generated hole in TiO_2, has already been mentioned and is effective with dialkyl sulfides [307]. Sulfides are known to quench triplet excited states of diaryl ketones by electron-transfer to produce sulfur radical cations [308, 309]. Thus irradiations of diaryl ketones such as 4-carboxybenzophenone, which is especially advantageous because of its water solubility for studying water soluble methionine and methionyl peptides, in the presence of sulfides generates sulfur radical cations by the sequence shown in Eqs. (44) and (45):

$$ArCOAr' \xrightarrow{hv} [ArCOAr']^*_{S_1} \xrightarrow{isc} [ArCOAr']_T \qquad (44)$$

$$[ArCOAr']^*_T + R_2S \longrightarrow [ArCOAr']^{\cdot-} + R_2S^{+} \qquad (45)$$

The diaryl ketone absorbs the light going to its singlet excited state which rapidly undergoes intersystem crossing to its excited triplet state. Electron transfer from the sulfide to the triplet excited state of benzophenone or 4-carboxybenzophenone is fast with rate constants of $(0.18-7)\times10^9$ $M^{-1}s^{-1}$ [308, 310–319] (for quenching rate constants with other diaryl ketone triplets and correlation of these rate constants with the free energy change for electron transfer see [314]). The sulfur radical cations produced in this way have been directly detected and their reactions monitored [310–319]. Intermolecular electron transfer from α-metalated sulfides to excited state DCA [320], phthalimide [321], acenaphthylenedione [322], and 2-cyclohexen-1-one [323] are known as is intramolecular electron transfer to a phthalimide [324], phenylglyoxylate [325], or 2-cyclohexen-1-one [323] or 1,8-naphthalimide [326] moiety. Irradiation of the charge transfer band of dialkyl sulfides with tetracyanoethylene [327] or with nitrogen dioxide [82] results in electron transfer from the sulfide to TCNE or NO^+, respectively. Photofragmentation of sulfonium peresters **136** as shown in Eq. (46) to give sulfur radical cations has been reported [328]:

$$R_2SCH(R')CH(R')CO_3t\text{-Bu} \xrightarrow{h\nu} R_2S^{+\cdot} + R'CH=CHR' + CO_2 + [t\text{-BuO}\cdot] \quad (46)$$
136

Direct spectroscopic evidence for the formation of sulfur radical cations was obtained in this reaction after laser flash photolysis.

Electrochemical methods have been used to generate sulfur radical cations but alkyl sulfides typically show irreversible oxidation [329–331]. The reason for this irreversibility is that the radical cation undergoes a rapid chemical reaction before it can be reduced back to the parent sulfide. The peak potentials that are measured in cyclic voltammetry typically are more positive than the reversible potential $E^{0\prime}$ because of slow heterogeneous electron transfer. However, cyclic voltammetry studies on 1,5-dithiocane proved to be uniquely insightful as will be discussed later in this review. Electrochemical peak potentials for oxidation of many dialkyl sulfides have been reported and tabulated. Correlations of these values with ionization potentials determined by photoelectron or charge transfer [332] spectroscopy have been made. However, there are important limitations to such correlations. Solvent effects [333] must be constant or negligible when comparing gas phase PES data with solution electrochemical data. The different time scales of the measurements may also have important consequences. The time scales of PES and CT spectroscopy are very different from the electrochemical time scale. Both PES and CT spectroscopy measure vertical ionization potentials, whereas, in an electrochemical experiment there is time for molecular reorganization involving atomic movement. An example believed to illustrate this difference has been reported [334]. The lowest ionization potential measured by PES for **137** and 1,5-dithiocane **138**, are 8.11 and 8.30 eV, respectively.

137   **138**   **139**   **140**

However, their anodic oxidation peak potentials in MeCN are 0.69 and 0.34 V vs Ag/0.1 mol l⁻¹ AgNO₃ in MeCN, respectively. That is, **138** is easier to oxidize electrochemically than **137** despite its higher ionization potential. Ab initio calculations support the notion that the $S\therefore S$ bond of **139** formed on oxidation of **138** is stronger than that for **137** after, but not before, geometry optimization.

Treatment of dialkyl sulfides with a variety of one-electron chemical oxidants has been suggested to give the corresponding radical cations. Because the radical cations are not stable under the reaction conditions, the evidence for their intermediacy is indirect. Consequently the generation of sulfur radical cations has not been unequivocally established for all of the reactions in which its formation has been invoked. Metal ion oxidants such as $SbCl_5$ [335], Cu(II) [336, 337], Tl(III) [338], tris (2,2′-bipyridyl) Fe(III) [339], and Ce(IV) [110, 340] have been used to effect one-electron oxidation of sulfides. In the case of oxidation of 1,5-dithiocane **138** with Cu(II) [336, 337] the corresponding radical cation **6** was unequivocally identified spectroscopically. Non-metal inorganic oxidants such as NO⁺ salts [336, 337], $O_2^{+\cdot}SbF_6^-$ [341], ·OH (generated by the reaction of Ti(III)

Table 10. EPR Spectroscopic parameters for dialkyl sulfide radical cations

Radical cation	Solvent	T, K	g_x	g_y	g_z	g_{av}	a, G
RR′S⁺							
R = R′ = Me	CFCl$_3$	130	2.0190	2.0145	2.0076	2.00137[a]	21 (6H)
	CFCl$_3$	77	2.033	2.016	2.001	2.017[b]	21.6, 21.4, 20.5
	CF$_3$CCl$_3$	81	2.019	2.019	2.0023	2.0134[c]	20.4 (6H)
R = R′ = CD$_3$	CFCl$_3$	77	2.032	2.015	2.002	2.0163[e]	3.3 (6D)
R = R′ = Et	CFCl$_3$	77	2.032	2.016	2.002	2.0170[d]	18–20 (2H)
R = R′ = i-Pr	CFCl$_3$	77	2.032		2.002	d	ca. 10 (2H)
R = R′ = t-Bu	CFCl$_3$	77	2.032	2.015	2.005	2.0160[d]	
	CFCl$_3$	77	2.033	2.017	2.006	2.019[b]	
	CH$_2$Cl$_2$	200				2.0130[e]	
R = CH$_3$, R′ = CH$_2$SiMe$_3$	CFCl$_3$	143				2.0145[f]	32.5 (^{33}S)
R,R′ = (CH$_2$)$_2$	CFCl$_3$	ca. 120	2.024	2.024	2.002	2.0166[d]	1.5 (3H), 12.5 (2H)
	CFCl$_3$	90	2.028	2.028	2.002	2.019[c]	31 (4H)
R,R′ = (CH$_2$)$_3$	CFCl$_3$	ca. 120	2.023	2.023	2.002	2.0160[d]	16.1 (14H)
	CFCl$_3$	90	2.027	2.027	2.003	2.019[c]	31 (4H)
R,R′ = (CH$_2$)$_4$	CFCl$_3$	77	2.027	2.014	2.002	2.0143[d]	31.1 (4H)
	CFCl$_3$	77	2.036	2.018	1.996	2.017[b]	20 (2H), 40 (2H)
R,R′ = (CH$_2$)$_2$SCH$_2$	CFCl$_3$	ca. 120				2.0080[d]	20.0
R,R′ = (CH$_2$)$_3$SCH$_2$	CFCl$_3$	77	2.018	ca. 2.008	2.002	d	ca. 6 (4 or 6H)
R,R′ = (CH$_2$)$_2$S(CH$_2$)$_2$	CFCl$_2$CF$_2$Cl	80	2.019	2.0092	2.002	d	≤6
	CFCl$_3$	130	2.0215	2.0158	2.0046	2.0118[g]	ca. 4.5
R,R′ = (CH$_2$)$_3$S(CH$_3$)$_3$	H$_2$SO$_4$	ca. 295	2.0194		2.0032	2.0128[g]	13, 3, 14, 3, 13,
	CH$_3$CN	ca. 263				2.012[h]	9.9 (4H)
	CH$_2$Cl$_2$	218				2.010[h]	11.2 (2H), 16.4 (2H)
R = R′ = Me$_2$N	CH$_3$NO$_2$	ca. 295				2.0053[e]	7.6 (2N), 7.6 (12H)
R = R′ = Et$_2$N	CH$_2$Cl$_2$	208				2.0050[i]	7.5 (2N), 7.5 (12H)
	CH$_3$NO$_2$	ca. 295				2.0058[e]	7.3 (2N), 5.0 (8H)
R = R′ = (CH$_2$)$_4$N	CH$_2$Cl$_2$	253				2.0056[i]	7.5 (2N), 5.0 (8H)
						2.0055[e]	7.6 (2N), 10.3 (8H)

R	Solvent	T (K)	g			a (hyperfine)
R=Me, R'=Me₂N	CH₃NO₂	ca. 295	2.0069[i]			12.1 (N), 8.7 (3H), 14.3 (3H), 14.4 (3H)
R=Me, R'=Et₂N	CH₃NO₂	ca. 295	2.0071[i]			14.3 (N), 8.6 (3H), 9.9 (2H), 8.6 (2H)
R=Me, R'=(CH₂)₄N	CH₂Cl₂	ca. 295	j			13.8 (N), 8.5 (3H), 21.7 (2H)
R=Me, R'=9–azabicyclo [3.3.1]nonane			j			14.1 (N), 8.3 (3H)
(RR'S)₂⁺·						
R=R'=Me	H₂O	ca. 295	2.0103[k]			6.8 (12H)
	c-C₃H₆	158	2.0102[e]			6.3 (12H)
			2.0103[l]			
R=R'=Et	H₂O	ca. 295	2.0104[m]			6.6 (12H)
	H₂O	ca. 295	2.0103[k]			6.7 (8H)
	H₂O	ca. 295	2.0103[m]			6.6 (8H)
	c-C₃H₆	165	2.0101[e]			6.3 (8H)
R=R'=i-Pr	c-C₃H₆	273	2.0112[e]			8.9 (4H)
R=t-Bu, R'=Me	c-C₃H₆	240	2.0107[e]			8.2 (6H)
R=t-Bu, R'=Et	c-C₃H₆	220	2.0108[e]			9.8 (4H), 32.1 (^{33}S)
R=t-Bu, R'=i-Pr	c-C₃H₆	220	2.0122[e]			6.0 (2H), 31.2 (^{33}S)
R=Me, R'=(CH₂)₂OH	H₂O	ca. 295	2.0102[k]			6.6 (10H)
R, R'=(CH₂)₂	CFCl₃	90	2.012[c]	2.0141	2.0122	2.0029	5.8 (8H)
R,R'=(CH₂)₃	CFCl₂CF₂Cl	100	2.00978[g]			5.0, 5.4, 3.6
	CFCl₂CF₂Cl	95	2.012[c]			9.4 (8H)
R,R'=(CH₂)₄	H₂O	295	2.0102[k,m]			9.3 (8H)
	c-C₃H₆	178	2.0100[g]			8.6 (8H)

a [350] c [352] e [354] g [356] i [358] k [342] m [361].
b [351] d [353] f [355] h [357] j [359] l [360]

with H_2O_2) [342], NH_3^+ (generated by the reaction of Ti(III) with NH_2OH) [343], and peroxynitrite [344] have also been employed. Both NO^+ salts and O_2^+ SbF_6^- have been used to produce **139**, identified spectroscopically, from 1,5-dithiocane. Oxidation of methionine and 2-keto-4-thiobutanoic acid with HOONO has been reported [344] to occur by two competing mechanisms. In the first, the corresponding sulfoxide is formed by a two-electron oxidation. In the second, ethylene is ultimately produced via an initial one-electron oxidation to give the corresponding sulfur radical cation. However, there was little one-electron oxidation of Thr-Met by HOONO [318]. The small amount of oxidation occurred by two-electron oxidation to the corresponding sulfoxide. Use of an organometallic one-electron oxidant, ferricenium hexafluorophosphate, has also been reported [345]. Oxidation of sulfides with triarylaminium (Ar_3N^+) salts [346] 2,3-dicyano-5,6-dichloroquinone [347], and N-fluoropyridinium tetrafluoroborate [348, 349] have also been reported to proceed by way of the corresponding radical cation.

2.1.2
EPR Spectroscopy

EPR Spectroscopic parameters for dialkyl sulfide radical cations are listed in Table 10. Such studies were crucial in unquivocally establishing the "dimeric" structure of dialkyl sulfide radical cations. Oxidation of a dilute aqueous solution of dimethyl sulfide with Ti(III) and H_2O_2, which generates ·OH, in a flow system provided the EPR spectrum of $(Me_2S)_2^+$ [342]. Eleven lines, with the correct relative intensities, could be resolved in this spectrum with the other two lines lost in the noise resulting from hyperfine splitting due to twelve equivalent hydrogen atoms. The EPR spectra of monomeric dialkyl sulfide radical cations could be measured in frozen matrices after γ-radiolysis. On annealing, or in more mobile matrices, these monomers associate with R_2S to form the 2c, 3e bonded $(R_2S)_2^+$ species. With cyclic dithioethers the 2c, 3e bonded species is observed in frozen matrices owing to the favorable intramolecular bond formation. Steric effects preclude t-Bu_2S^+ from association, so that the EPR spectrum of this species can be measured in solution. However, in other cases the associated $(R_2S)_2^+$ species is invariably observed. Conformational behavior of these species from cyclic sulfides can be ascertained by EPR spectroscopic analysis. Of particular interest in this regard is 1,5-dithiocane radical cation **139**. The EPR spectrum of this species in acetonitrile consists of a triplet of triplets with a_H=16.4±0.1 G (2H) and a_H=11.2±0.1 G (2H) [357]. There are two pairs of equivalent protons interacting with the unpaired electron of the eight β-hydrogens. The hyperfine splittings depend on $\cos^2 \theta$ where θ is the dihedral angle between the β-C-H bond S-S internuclear axis. It is estimated that these angles are 32° and 45° to account for a_H of 16.4 and 11.2 G, respectively. The dihedral angles between the remaining β-C-H bonds would result in small <3 G splitting. A structure consistent with this analysis is **140** in which there is a transannular S-S bond forming two cis-fused five-membered rings in an eight-membered ring with an overall boat chair conformation devoid of conformational mobility equilibrating both five-membered rings. Resonance Raman spectroscopic studies on this radical cation provided further evidence for a transannular S-S bond [362]. Such studies provided a tentative assignment of the 2c, 3e S,S stretching frequency to a prominent band at 118 cm^{-1}.

2.1.3
Two-Center Three-Electron Bonds

The propensity of dialkyl sulfide radical cations to bond to electron-rich centers exemplified by reaction with parent dialkyl sulfide to form a 2c, 3e S, S bond as shown in Eq. (47) has attracted considerable attention:

$$R_2S^{+\cdot} + R_2'S \rightleftharpoons (R_2SSR_2')^{+\cdot} \qquad (47)$$
$$\mathbf{141}$$

The nature of a 2c, 3e bond has already been discussed in this review but the equilibrium constant for its formation and its bond strength needs to be addressed. Gas phase studies using mass spectroscopic techniques and theoretical computations have been reported for Eq. (47) where, R=R'=Me [363, 364], R=R'=Et [364, 365] and R=Me, R'=Et [365]. The experimentally determined ΔH^0 values and the calculated values for the 2c, 3e S,S bond are listed in Table 11. Calculations of the bond enthalpies at 298 K show that the relative bond strength are **141**, R=R'=Me>**141**, R=R'=Et>**141**, R=Me, R=Et [365]. The lower bond energy for **141**, R=Me, R=Et is expected on the basis of ΔIP as outlined above. Furthermore, the bond strengths are roughly 45% of the S-S single bond energy which agrees with the idea that the formal bond order is 1/2. For equilibrium **141**, R=R'=Me, a ΔG^0 of –56.1 [363] and –52.2 kJ/m [364] was measured at 525 K in the gas phase. In aqueous solution the equilibrium constant for this reaction is reported to be $(2.0\pm0.3)\times10^5$ M^{-1} [366], and 2.5×10^4 M^{-1}. Thus the ΔG^0 is much less favorable in aqueous solution than in the gas phase. It has been suggested [142, 367] that a 2c, 3e S,O bond forms between Me$_2$S$^+$ and H$_2$O. Consequently, in aqueous solution Eq. (48) better represents the equilibrium measured:

$$(Me_2SOH_2)^{+\cdot}_{solv} + Me_2S \rightleftharpoons (Me_2S)^{+\cdot}_{2,\,solv} + H_2O \qquad (48)$$

This would account for the discrepancy between aqueous solution and gas phase data. The role of water on equilibrium (Eq. 49) was addressed computationally [364].

$$H_2S \cdot 2H_2O + H_2S^{+\cdot}\, 2H_2O \longrightarrow [H_2S \therefore SH_2]^{+\cdot}\, 4H_2O \qquad (49)$$

It was found that the 2c, 3e S,S bond energy remains almost the same in water as in the gas phase suggesting that the solvation energy of H$_2$S$^+$ and (H$_2$S $\therefore$ SH$_2$)$^+$ are nearly the same. However, hydrogen bonding in which (H$_2$S $\therefore$ SH$_2$)$^+$ acts as the donor and 4H$_2$O as the acceptors was found which is not possible for (Me$_2$S)$_2^{+\cdot}$. A correlation between the bond strength in 2c, 3e S,S bonds and λ_{max} for these species was suggested based on the assignment of this absorption to a

Table 11. Experimental and theoretical values for ΔH^0 of 2c, 3e S,S bonds

Radical cation	ΔH^0 calcd kJ/m	ΔH^0 expt kJ/m
$(Me_2S)^{\ddagger}$	109.6[a]	
	122[b]	111±2[b]
$(Et_2S)^{\ddagger}$	112[c]	
	121.3[d]	119±1.5[d]
	118.8[e]	123[e]
$(Me_2SSEt_2)^{\ddagger}$	107.8[h]	104±1.5[f]
		107[g]

[a] Ab initio calculations at the [PMP2/6–31G/d)]//3–21G (d) level [363].
[b] At 576 K, ab initio calculations at the [PMP4/6–31+G(2df,p)]//MP2/6–31G (d) level [364].
[c] At 520 K, ab initio calculations at the [PMP]2/6–31G(d)]//HF/6–31G (d) level [364].
[d] At 506 K, DFT calculations at the B3LYPL/6–31G(d) level [365].
[e] At 0 K, experimental ΔH^0 for bond breaking corrected to 0 K, DFT calculations at the B3LYP/6–31G(d) + ZPC level [365].
[f] At 506 K [365].
[g] At 0 K, experimental ΔH^0 for bond breaking corrected to 0 K using ab initio parameters, DFT calculations at the B3LYP/6–31(d)//B3LYP/6–31G(d) + ZPC level [365].

$\sigma \rightarrow \sigma^*$ transition [368]. In addition a linear correlation has been reported [369] between λ_{max} and the Taft substituent parameter σ^* for unbranched substituents [369, 370]. However, this absorption is due to a σ-lone pair $\rightarrow \sigma^*$ transition calling these correlations into question [364]. Nevertheless it has been shown [368–370] that substituents and conformational constraints have a dramatic effect on λ_{max} of S∴S bonded radical cations and their stability. With regard to the stability of these species, it has been cogently argued [368, 370, 371] that the strength of the S∴S bond formed intramolecularly on one-electron oxidation of dithioethers depends on the S-S distance and angle between sulfur p-orbitals. These geometric parameters determine the extent of overlap of the interacting atomic orbitals and, in turn, reflect conformational constraints. This is well-illustrated by the remarkable radical cation **139** which is unusually stable. An acetonitrile solution of this species as its tetrafluoroborate salt is stable at room temperature for several days [336, 337]. Its stability is undoubtedly due to the favorable intramolecular formation of two five-membered rings in which the S-S distance and p-orbital overlap is optimal. Another example which illustrates the influence of substituents is t-Bu$_2$S‡. Unlike R$_2$S‡ species in general, which as already pointed out typically dimerize, t-Bu$_2$S‡ (λ_{max}=310 nm) shows no tendency to dimerize in aqueous solution [369]. This effect is ascribed to steric inhibition of dimerization by the bulky $tert$-butyl groups. This effect may also contribute to the diminished value of K for equilibrium (Eq. 47) where R=R'=i-Pr of 5.4×10^2 M^{-1} compared with Eq. (47), R=R'=Me, of 2×10^5 M^{-1} in aqueous solution [366]. However, inductive effects also contribute. Clearly the above delineated factors play an important role in S∴S bond strength qualitatively tracked by λ_{max}. However, λ_{max} may prove unsuitable for quantitative comparisons of bond strength because factors may affect the energy of the σ-lone pair orbital, thereby affecting λ_{max} which, as pointed out above, involves the σ-lone pair $\rightarrow \sigma^*$ transition. But since the σ-lone pair orbital is not involved in $S \therefore S$ bond, it would not affect $S \therefore S$ bond strength. Specifically, alkyl groups by symmetry can interact more with the σ-lone pair orbital than the σ^* orbital. Consequently, the destabilized σ-lone pair orbital moves closer in energy to σ^* resulting in a red-shift in λ_{max} [142]. In addition, calculations on the effect of water solvation on $(H_2S)_2^+$ predict a blue-shift in λ_{max} [364]. Both of these effects can occur without significantly affecting the $S \therefore S$ bond strength. Thus the determination of the quantitative dependence of $S \therefore S$ bond strength on substituent structure must await further studies.

As pointed out above a 2c, 3e bond between Me_2S^+ and H_2O has been suggested. Analogous bonding may occur between alcohols and sulfur radical cations. Such bonding is favored in intramolecular cases in which a five- or six-membered ring forms and especially in systems such as **142** in which addition conformational constraints favor S,O bond formation.

MeS CMe_2OH MeS $\overset{..}{\underset{+ \, :O}{}}$ CMe_2 MeS $\overset{..}{\underset{:O}{}}$ CMe_2

142   **143**   **144**

Radical cation **143** is formed by oxidation of **142** with ·OH under pulse radiolysis conditions in water and identified by its absorption spectrum with λ_{max}=420 nm [372]. A similar species has been obtained under pulse radiolysis conditions from $S(CH_2CH_2CH_2OH)_2$ and it absorbs at 410 nm [294, 298, 372]. The pK_a of **143** is 5.9 and its conjugate base **144** absorbs at λ_{max}=400 nm. Such neutral radicals can also be obtained by bond formation between sulfur radical cations and carboxylate moieties. Thus oxidation of **145** under pulse radiolysis conditions gave transient **146a** with λ_{max}=390 nm.

MeS CO_2^- MeS $\overset{..}{\underset{:O}{}}$ S   NMe

145   **146a**, R=H   **147**
                     b, R=NH_3^+

Similarly oxidation or photoelectron transfer of $MeS(CH_2)_3CO_2H$, $MeS(CH_2)_2CO_2H$, and $S(CH_2CH_2CO_2H)_2$ gave analogous species with λ_{max} around 400 nm [372–374] and **146b** absorbs with λ_{max}=340 nm [375]. Although these species were considered to involve 2c, 3e S,O bonds they are analogous to **54** previously discussed for which 3c, 3e bonding was suggested. Consequently the bonding in these systems needs to be further elucidated.

Oxidation of **147** with $Br_2^{\cdot-}$ under pulse radiolysis conditions generates the corresponding radical cation with an $S \therefore N$ bond [376] and which absorbs with a broad maximum at 500 nm. This species could also be produced by reaction of the corresponding dication with **147** [377] and shows a broad signal in its EPR spectrum with $g_{av}=2.02$. Oxidation of $MeS(CH_2)_3NH_2$ with $\cdot OH$ under pulse radiolysis conditions produced the corresponding radical cation with an intramolecular $S \therefore N$ bond which absorbed with $\lambda_{max}=385$ nm [374]. Methionine **148a** undergoes oxidation under these conditions to give a transient with $\lambda_{max}=400$ nm [374].

148a, R=CO_2^- **149a**, R=CO_2^- **150a**, R=CO_2^-, R'=NH_3^+
   b, R=$CONHCH_2CO_2^-$ **b,** R=$CONHCH_2CO_2^-$ **b,** R=NH_3^+, R'=CO_2^-
   c, R=$CONHCH_2CONHCH_2CO_2^-$ **c,** R=$CONHCH_2CONHCH_2CO_2^-$

In principle, this oxidation product could have an $S \therefore N$ or $S \therefore O$ bond. The similarity of the absorption of this product and that obtained by oxidation of $MeS(CH_2)_3NH_2$ and methionine ethyl ester suggested an $S \therefore N$ bond as shown in **149a**. This suggestion was unequivocally proven by oxidation of constrained methionine analogues. As already pointed out, **146b** which has an $S \therefore O$ bond absorbs with λ_{max} at 340 nm. However, oxidation of **150a** [378] under pulse radiolysis conditions gave a one-electron oxidation product which can only form an $S \therefore N$ bond intramolecularly and it absorbs with $\lambda_{max}=400$ nm [204], which matches the absorption of the one-electron oxidation product of methionine whereas the absorption of **146b** does not. Oxidation of methionine by photoelectron transfer produced the same species as that formed under pulse radiolysis conditions [315]. As the concentration of methionine is increased $S \therefore S$ bonded dimers are formed at the expense of $S \therefore N$ species on oxidation, providing evidence for the greater thermodynamic stability of $S \therefore S$ bonds over $S \therefore N$ bonds. Oxidation of methionyl di- and tripeptides also results in the formation of $S \therefore N$ bonded species, provided that the methionine amino group is not acylated, so that formation of the $S \therefore N$ bond results in a five-membered ring. Consequently $S \therefore N$ bonded species are formed on oxidation of Met-Gly, **148b**, but not Gly-Met [317]. Thus, under pulse radiolysis conditions using $\cdot OH$ or photoelectron transfer conditions using 4-carboxybenzophenone Met-Gly gives $S \therefore N$ bonded transient **149b** absorbing at $\lambda_{max}=380$ nm and Met-Gly-Gly (but not Gly-Gly-Met nor Gly-Met-Gly) similarly yields the $S \therefore N$ bonded species **149c** with $\lambda_{max}=380$ nm. With all of these species increasing the concentration of di- or tripeptide results in formation of intermolecularly $S \therefore S$ bonded dimers with λ_{max} 480–490 nm. Recently high level calculations on $S \therefore N$ bonded radical cations have been reported [379].

Two-centered, three-electron bonds between R_2S^+ and halide ions (Cl^-, Br^-, and I^-) have been   identified [369, 380]. The K values for the dissociation equilibrium (Eq. 50) determined by pulse radiolysis techniques in water are 1.2×10^{-2}, 8.2×10^{-6}, and $\ll 10^{-6}$ for X=Cl, Br, and I, respectively.

$$Me_2S \therefore X \rightleftharpoons Me_2S^{\overset{+}{\cdot}} + X^- \qquad (50)$$
$$\mathbf{151}$$

The λ_{max} values reported for **151**, X=Cl, Br, and I are 390, 400, and 410 nm, respectively. There is another reason for interest in these species. As already mentioned, oxidation of sulfides to the corresponding radical cation under pulse radiolysis conditions is sometimes accomplished using $Cl_2^{\overline{\cdot}}$, $Br_2^{\overline{\cdot}}$, or $I_2^{\overline{\cdot}}$ as oxidant. The intermediates in these oxidations are **151**. Alkyl bromides and iodides but not chlorides can also participate in S∴X bonding at least intramolecularly [369, 381]. Oxidation of **152**, X=Br or I with ·OH, under pulse radiolysis conditions forms **153**, X=Br or I, identified by their absorption spectrum. As the concentration of **152** is increased and, in all cases for **152**, X=Cl, the spectrum of the intermolecularly S∴S bonded species is observed.

Oxidation of **154**, n=0, 1, and 2 by ·OH under pulse radiolysis conditions generates a transient which absorbs at λ_{max}=480, 385, and 440 nm, respectively [382, 383]. Conductivity measurements indicate that a radical cation is formed in each case.

$$RS(CH_2)_nX \qquad\qquad RS \overset{\overset{\displaystyle (CH_2)_n}{\frown}}{\underset{+}{\therefore}} X$$
$$\mathbf{152} \qquad\qquad\qquad \mathbf{153}$$

$$MeS \quad PEt_2 \qquad\qquad MeS \underset{+}{\therefore} PEt_2$$
$$\mathbf{154} \qquad\qquad\qquad \mathbf{155}$$

These intermediates have been assigned the corresponding structure **155**, n=0, 1 and 2, respectively, with an S∴P bond.

The evidence for 2c, 3e S∴S, S∴O, S∴N, S∴X, and S∴P bonds is reviewed above. In addition, EPR spectroscopic evidence for S∴S, S∴N, S∴Cl, and S∴Br bonds involving methionine and some of its derivatives and an analogue have been reported [384]. The species were formed under pulse radiolysis conditions in water. The EPR spectra of these species were measured and their UV/VIS absorption spectra correlated them with the species previously studied by pulse radiolysis techniques. The EPR spectroscopic data provided further convincing evidence for the assigned structures because the hyperfine splitting constants for the bonded Cl, N, and Br nuclei and g values are diagnostic for these species. The relative stabilities of these species was found to follow the order of increasing stability: S∴O<S∴Cl<S∴Br<S∴N<S∴S, which is as expected based on the ΔIP criterion for 2c, 3e bond strength.

In all of the preceding cases a nonbonding pair of electrons is used to stabilize a sulfur radical cation by forming a 2c, 3e bond. It has also been reported [385] that π-electron donors can stabilize sulfur radical cations. Monomeric or dimeric radical cations were generated from thiirane, thietane, thiolane, thiane, and dimethyl sulfide in toluene although higher concentrations of sulfide were required for dimer formation in toluene than in alkanes. The EPR parameters for these species in toluene were determined by time-resolved fluorescence-detected magnetic resonance [386]. The EPR parameters for the dimers in toluene were comparable to those obtained from the EPR spectrum measured in $CFCl_3$ reported above in Table 10. However, the a_H values and g factors for the monomeric radical cations in toluene are smaller than those in $CFCl_3$. The coupling constants are about 1/3 but the value depends on thioether ionization potential. These results suggest complexation of the radical cation by toluene (and xylene as well) despite the lack of observed hyperfine interactions with toluene but perhaps more than one arene molecule is involved thereby decreasing the splitting constant. The nature of the interaction is not clearly defined but π-complexation of $R_2S^{\ddagger}$ has been suggested.

Electrochemical studies [387, 388] on 1,5-dithiocane **138** proved uniquely insightful because this unusual sulfide, in contrast to ordinary sulfides, shows reversible behavior. This reversibility is due, in part, to the stability of **139** owing to its transannular S∴S bond as outlined above. At low concentrations in acetonitrile as solvent **138** undergoes two reversible one-electron oxidations as shown in Eqs. (51) and (52):

$$138 \rightleftharpoons 139 + e^- \ E_1^{0'} \tag{51}$$

$$139 \rightleftharpoons 156 + e^- \ E_2^{0'} \tag{52}$$

In the first step **138** is oxidized to **139** and, in the second step, **139** is oxidized to dication **156**. However, only one oxidation peak is observed with E_p=0.34 V vs Ag/0.1 mol l^{-1} AgNO$_3$ in CH$_3$CN which is less positive than that of ordinary sulfides (E_p=1.2–1.7 V). Thus **138** is reversibly oxidized exceptionally easily. Only one oxidation peak is observed because $E_2^{0'}$ is slightly less positive than $E_1^{0'}$ by ca. 20 mV. On the basis of coulombic interactions $E_1^{0'}$ should be less positive than $E_2^{0'}$ but the opposite is seen. An electronic reason for this result is suggested by the 2c, 3e bonding scheme outlined above. That is, the electronic configuration of the three electrons shared between the two sulfur atoms is $\sigma^2\sigma^{*1}$. Consequently, oxidation of **139** results in the removal of an antibonding electron resulting in increased S-S bonding. Apparently this electronic effect is energetically more important than the Coulombic effect. Thus the electrochemical studies showed two chemical consequences of transannular S∴S bond formation in oxidation of **138**: low oxidation potentials for both **138** and **139**.

156   157

Dication **156**, produced by the oxidation of **138** is also relatively stable. Furthermore because $E_2^{0'}$ is less than $E_1^{0'}$ disproportionation of **139** to give **138** and **156** is favorable in contrast to the unfavorable disproportionation of TH‡. Consequently, the mechanism for nucleophilic attack on **139** is likely to involve the disproportionation mechanism outlined above for $TH^{\ddagger}$. It is, therefore, relevant to discuss briefly dication **156**. This dication was first prepared by oxidation of NO^+. It can also be prepared by treatment of sulfoxide **157** with H_2SO_4 [389, 390] or with $(CF_3SO_2)_2O$ [391]. In the latter case, crystals of **156** $(CF_3SO_3^-)_2$ can be obtained and the structure of this species has been determined by X-ray crystallographic methods [392]. The eight-membered ring adopts a chair-chair conformation and the S-S bond length is 2.124 Å in the solid state. There are close contacts between the S atoms of the dication and the O atoms of the anions. Dication **156** reacts with water to give **157**. A kinetic study on the reduction of **157** by HI to **138** has shown that dication **156** is first formed in the rate determining step [393]. Dication **156** is then attacked by iodide ion to give the corresponding sulfonium iodide which on further reaction with I^- gives **138** and I_2. Dication **156** also effects electrophilic aromatic substitution with aniline, triphenyl amine and phenol but redox reactions with thiophenol and 1,2-diphenylhydrazine [394].

2.1.4
Electron Transfer

As outlined above, pulse radiolysis methods were used to determine $E^0=1.66\pm0.03$ V vs SHE for Eq. (53) R=Me, $E^0=1.63$ V vs SHE for Eq. (53) R=t-Bu, and $E^0=1.40\pm0.02$ V vs SHE for Eq. (54) [289].

$$R_2S-e^- \rightleftharpoons R_2S^{\ddagger} \tag{53}$$

$$2Me_2S-e^- \rightleftharpoons (Me_2S)_2^{\ddagger} \tag{54}$$

Consequently, these species are good one-electron oxidants and have often been used as such in pulse radiolysis experiments. For example, $Me_2S^{\ddagger}$ oxidizes MeSSMe to $(MeSSMe)^{\ddagger}$ with a bimolecular rate constant of $(4.0\pm0.4)\times10^9$ $M^{-1}s^{-1}$ and t-$Bu_2S^{\ddagger}$ oxidizes EtSSEt to $(EtSSEt)^{\ddagger}$ with a bimolecular rate constant of 5×10^9 $M^{-1}s^{-1}$ [395, 396]. Oxidation of $Fe(CN)_6^{4-}$ to $Fe(CN)_6^{3-}$ by $(Me_2S)_2^{\ddagger}$, $(Et_2S)_2^{\ddagger}$, or $[(CH_2)_4S]_2^{\ddagger}$ is essentially diffusion controlled [291]. The oxidation of thiols by t-$Bu_2S^{\ddagger}$ and $Me_2S^{\ddagger}$ has been studied [397] and the rates for these reactions are close to diffusion controlled. Although the rate of oxidation of thiolates by $(Me_2S)_2^{\ddagger}$ is also close to the diffusion limit, oxidation of thiols by this species is slower.

2.1.5
Attack by Nucleophiles

As pointed out above, nucleophiles such as sulfides, alcohols, amines, phosphines, halides, and halide ions readily attack sulfur radical cations forming $S\therefore S$, $S\therefore O$, $S\therefore N$, $S\therefore P$, and $S\therefore X$ species. Generation of $Me_2S^{\ddagger}$, under pulse radiolysis conditions, by the reaction of Me_2SO with $H\cdot$ in aqueous $HClO_4$ permitted the measurement of rate constants for nucleophilic attack on this species. The rate constants for $Me_2S^{\dot{+}} + Me_2S \rightarrow (Me_2S)_2^{\dot{+}}$ and for $Me_2S^{\dot{+}} + t\text{-}Bu_2S \rightarrow (Me_2SSt\text{-}Bu_2)^{\dot{+}}$ determined in this way are $(3.0 \pm 0.3) \times 10^9$ M^{-1}s^{-1} and $(4.0 \pm 0.4) \times 10^9$ M^{-1}s^{-1}, respectively [396]. In the presence of Cl$^-$ nucleophilic attack by this species was also observed to form $Me_2S \therefore Cl$ identified by its absorption with $\lambda_{max} = 380$ nm. Nucleophilic attack by $^-$OH on $Me_2S^{\dot{+}}$ [289] to give $Me_2S\text{-}OH$ (discussed above) and on **139** to form **158a** with $\lambda_{max} = 370$ nm [311] have been observed spectroscopically.

$$R_2S\text{---}\dot{S}R_2$$
$$|$$
$$OH$$

158a, X = OH   **159**
     b, X = I

Nucleophilic attack on **139** by I$^-$ produces **158b** shown spectroscopically, $\lambda_{max} = 370$ nm [311].

Oxidation of Me_2S, Bu_2S, and thiolane by O_2 is catalyzed by NO_2 [82]. Furthermore, irradiation of the NO_2 or $NO^+BF_4^-$ complex of Me_2S or thiolane generates the corresponding sulfur radical cation as shown by time resolved spectroscopy. In analogy with the mechanism suggested above for such oxidation of TH, NO_3^- attacks the sulfur radical cations to produce ultimately the corresponding sulfoxides.

Photooxidation of dialkyl sulfide sensitized by DMN in polar solvents gives the corresponding sulfoxides [131]. As discussed above for such oxidations of Ph_2S, $O_2^{\dot{-}}$ reacts with $R_2S^{\dot{+}}$.

Pulse radiolysis studies on the oxidation of Me_2S and Et_2S by $\cdot$OH in the presence of O_2 have shown that the dimeric radical cation is attacked by $^-$OH to give **159** or $R_2\dot{S}OH$ and R_2S [398]. The subsequent reactions of these species are delineated below in the discussion of oxidation reactions. The rate of reaction of $(Me_2S)^{\dot{+}}$ with $O_2^{\dot{-}}$ is extremely fast and the rate constant for this reaction has been found to be $(2.3 \pm 1.2) \times 10^{11}$ M^{-1}s^{-1} [399].

2.1.6
α-Deprotonation

α-Deprotonation of sulfur radical cations is a major reaction pathway for the decomposition of these species. The product of deprotonation $\ce{>C}$ – SR absorbs with $\lambda_{max} = 280$ nm which permits the easy monitoring of its formation under pulse radiolysis conditions. This species can also be detected by EPR spectroscopy in flow systems [343]. The pK_a of $Me_2S^{\dot{+}}$ has been estimated [289] as 0 but there is apparently a kinetic barrier to deprotonation as there is for other carbon acids. Deprotonation of $R_2S^{\dot{+}}$ occurs much more readily than deprotonation of $(R_2S)_2^{\dot{+}}$ [366]. The consequences of this difference are that increasing the concentration of R_2S increases the lifetime of the radical cation and $(Et_2S)_2^{\dot{+}}$ is less stable than $(Me_2S)_2^{\dot{+}}$ because of a 10× faster deprotonation of $Et_2S^{\dot{+}}$ [291].

α-Deprotonation of the radical cations of 1,3-dithiolanes **160** has been studied [400]. Preferential deprotonation at C-2 of **160a**‡ is expected.

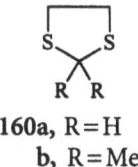

160a, R=H
b, R=Me

The increased stability of **160b**‡ over **160a**‡ is in accord with such expectations but another factor may be responsible for this result. It is suggested [400] that the intramolecular S∴S bond in **160b**‡ is stronger than that in **160a**‡ owing to geometry differences which result in better orbital overlap for **160b**‡ than **160a**‡. Anodic oxidation of alkylthioacetonitriles [401], $RSCH_2CN$, in methanol in the presence of $TsOH \cdot H_2O$ results in deprotonation of the radical cation produced. The radical RSĊHCN thus formed is further oxidized to RS^+CHCN which is attacked by methanol. The yields of the resulting products RSCH(OMe)CN for R=Me, Et, and t-Bu are 70, 86, and 91%, respectively. Products from deprotonation of sulfur radical cations formed by oxidation of thioethers with N-fluoro-pyridinium tetrafluoroborate have also been identified [348]. In addition to their formation by α-deprotonation, $>\!\dot{C}-SR$ can be formed directly by hydrogen atom abstraction. Analysis of the formation of these radicals under pulse radiolysis conditions is aided by their reducing properties especially with $C(NO_2)_4$ to generate the strongly absorbing $C(NO_2)_3^-$, $\lambda_{max}=350$ nm [402]. The maximum yield of $>\!\dot{C}-SR$ obtained by direct hydrogen atom abstraction is ca. 20% of the total $\cdot OH$ radicals available [402].

2.1.7
Reaction with O$_2$

Under pulse radiolysis conditions $(R_2S)_2^{\ddagger}$ and $R_2S^{\ddagger}$ do not react with O_2, that is, $k \le 10^6$. However, the yield of Me_2SO formed from the reaction of $\cdot OH$ and Me_2S in aqueous solution is greatly increased in the presence of O_2 [398]. The reason for this is that **159** reacts readily with O_2 to form $MeS(OH)S(O_2\cdot)Me$ which decomposes to Me_2SO, Me_2S and O_2 or **159** reacts with O_2 to give $Me_2S(OH)O_2\cdot$ and Me_2S. Decomposition of $Me_2S(OH)O_2\cdot$ yields Me_2SO and O_2^-. This same intermediate $MeS(OH)O_2\cdot$ can be formed in an alternative way. Reaction of MeSOH with O_2 gives this species and the rate constant for this reaction is 2×10^8 $M^{-1}s^{-1}$ [289]. The rate constants for the reactions of adducts **134** and **135** are $(1.13\pm0.07)\times10^8$ and $(1.28\pm0.10)\times10^8$ $M^{-1}s^{-1}$, respectively [293]. Similarly adduct **158a** reacts with O_2 to form the corresponding sulfoxide, H^+ and O_2^- [311]. Photooxidation of Me_2S and Et_2S using triplet 4-carboxybenzophenone and O_2 has been studied [403]. As already pointed out, triplet diaryl ketones undergo electron transfer with sulfides to generate sulfur radical cations and Ar_2CO^-. In the presence of O_2, Ar_2CO^- forms O_2^- by electron transfer. It is found that $Me_2S^{\ddagger}$ reacts, under these conditions, mainly but not entirely by a pathway that is inhibited by superoxide dismutase. The superoxide dismutase inhibited pathway involves the reaction of $(Me_2S)_2^{+}$ with O_2^{-}, as reported above to form $Me_2S^+OO^-$ and Me_2S. Zwitterion $Me_2S^+OO^-$ is the same species formed by the reaction of $^1O_2$ with Me_2S and is known to react further with Me_2S to give $2Me_2SO$.

It has also been reported [61] that ammonium cerium(IV) nitrate catalyzes the oxidation of R_2S by O_2 to give R_2SO. The mechanism suggested for this reaction is as follows. Cerium(IV) oxidizes R_2S to give R_2S^{+} which reacts with O_2 to form the radical coupling product $R_2S^+OO\cdot$. This species then reacts with Ce(III) to produce $R_2S^+OO^-$, which as stated above reacts with R_2S to give $2R_2SO$, and Ce(IV), which renders the sequence catalytic in Ce(IV). Although it was pointed out above that no reaction was observed between R_2S^{+} and O_2 under pulse radiolysis conditions in water, the reactions catalyzed by Ce(IV) were run under different conditions with high O_2 pressures and in acetonitrile or aqueous acetonitrile as solvent.

2.1.8
Decarboxylation

Photooxidation of methionine **148a** with diaryl ketones results in the formation of the corresponding radical cation as does oxidation with $\cdot OH$ under pulse radiolysis conditions. Electron transfer from N to S generates the corresponding nitrogen radical cation which decarboxylates to $R\dot{C}HNH_2$ [308, 404]. If the amino group of Met [126] is acylated or with conformationally constrained Met analog **150b** [375], in which S···N interaction is precluded, decarboxylation still occurs. In these cases electron transfer from the carboxylate moiety to the sulfur radical cation is suggested to occur. The resulting acyloxy radical decarboxylates. In the case of γ-Glu-Met decarboxylation of the Met residue occurs as expected on $\cdot OH$ oxidation but decarboxylation of the γ-Glu residue happens as well [406]. Interaction between the γ-GluNH$_3^+$ group and $R_2\dot{S}OH$ is suggested [406] despite their separation by seven bonds. Similar α-decarboxylation of the γ-Glu residue occurs on $\cdot OH$ oxidation of S-alkylglutathione derivatives.

2.1.9
C-S Bond Cleavage

Anodic oxidation of 4-methoxybenzyl thioethers mediated by $(4\text{-}BrC_6H_4)_3N^{+}$ results in C-S bond cleavage [407]. The use of the 4-methoxybenzyl group as a protecting group for cysteine in peptide synthesis is recommended because this protecting group can be selectively deblocked to form the disulfide cystine in high yields. Oxidation of substituted thiiranes with 2,3-dichloro-5,6-dicyano-p-benzoquinone [408] or Ce(IV) [409] in alcohols or acetic acid provides alkoxy or acyloxydisulfides **161**. This reaction is suggested to proceed via one-electron oxidation of the thiirane followed by nucleophilic attack with C-S cleavage. In non-nucleophilic solvents oxidation of tri- and tetraaryl thiiranes with $(4\text{-}BrC_6H_4)_3N^{+}$ yields the corresponding alkene in excellent yield in which both C-S bonds are cleaved [410]. Cleavage of the C-S bond in dialkyl sulfur radical cations is especially useful with monothioacetals. As already discussed above with 1-phenylthioglycosides, electrochemical oxidation of 1-ethylthioglycosides [175] results in C-S cleavage in the radical cation. The resulting glycosyl cation then reacts with alcohols including sugar alcohols to generate glycosides and disaccharides. Anodic oxidation of methylthiomethyl ethers **162a** (protected alcohols) in NaOAc, AcOH produces **162b** which, on treatment with base affords alcohols RCH$_2$OH in excellent yields [411].

RCHCH$_2$S$\frac{}{}_2$
|
OR'

161

RCH$_2$OCH$_2$X

162a, X = SMe
b, X = OAc

OMe
|
CH$_2$CHSR
|
O $\sim$ N $\sim$ O

163a, R = Me
b, R = t-Bu

Irradiation of **163a** or **b** generates the corresponding sulfur radical cation which undergoes C-S cleavage. This results in the formation of the corresponding cation which reacts with a variety of nucleophiles such as MeOH, EtOH, and t-BuNH$_2$ [326]. Of special interest are the DNA cleaving properties of **163a** on irradiation due to alkylation of DNA and subsequent facile hydrolysis. The photogenerated cation alkylates N-benzoyl-2'-deoxyadenosine at N7 among other products and such N7 alkylation of adenine residues in DNA was shown to occur.

2.1.10
α C-M Bond Cleavage

As already pointed out, anodic oxidation of α-stannyl and α-silyl phenyl thioethers is known to result in C-M bond cleavage to generate PhSC$\overset{+}{\underset{.}{}}$ which in turn reacts with nucleophiles. This overall sequence is synthetically useful. Similarly α-metalated dialkyl sulfides undergo C-M bond cleavage following one-electron oxidation. Photoelectron transfer from α-metalated sulfides to a variety of acceptors has already been mentioned as a method for generating sulfur radical cations. These radical cations undergo overall carbon-carbon bond forming reactions to give demetalated products. The mechanisms for these reactions most likely involve metal transfer from the α-metalated sulfur radical cation to the acceptor radical anion, then radical-radical coupling, and finally protonolysis of the metal on workup. Thus irradiation of DCA and Me$_3$SiCH$_2$SEt affords **164** in 90% yield [320]. Irradiation of phthalimide [321], N-methylphthalimide [321] or acenaphthylenedione [322] provide **165a**, **165b**, and **166a** in 85, 78, and 20% yields, respectively. Interestingly, the irradiation of acenaphthylenedione and Me$_3$SiCH$_2$SEt was carried out in protic solvent MeOH but, if this reaction is done in aprotic solvent MeCN, proton transfer to give **166b** in 19% yield occurs as well. Irradiation of MeSCH$_2$SnBu$_3$ or (MeS)$_2$CHSnBu$_3$ and 2-cyclohexen-1-one produces **167a** or **b** in 50 and 41% yields, respectively [323]. Intramolecular analogues of this intermolecular reaction work well.

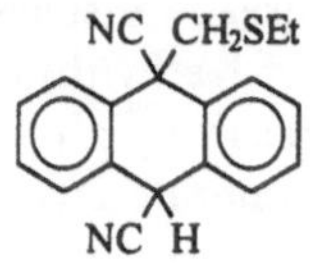

NC CH₂SEt

164

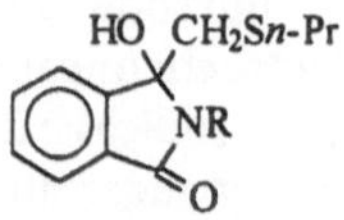

HO CH₂Sn-Pr

NR
O

165a, R=H
b, R=Me

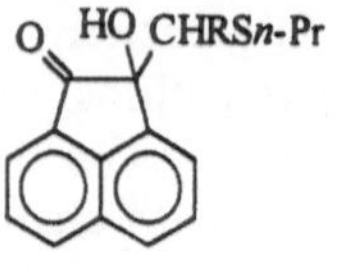

O HO CHRSn-Pr

166a, R=H
b, R=Me₃Si

O

CHRR'

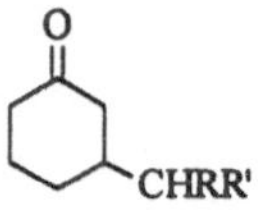

167a, R=SMe, R'=H
b, R=R'=SMe

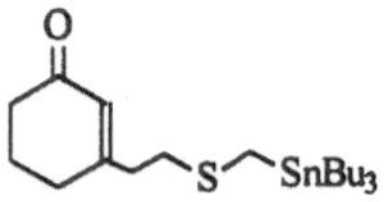

O

S SnBu₃

168

O

Ph

Bu₃ Sn

S

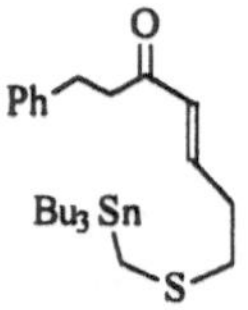

169

Thus irradiation of **168** or **169** in the presence of 1,4-naphthalenedicarbonitrile gave **170** and **171** in 84 and 61 % yields, respectively [323]. The mechanism for these reactions is analogous to that suggested for the preceding examples but the yields for the α-silylated sulfides proved far inferior to those of the corresponding α-stannyl sulfides.

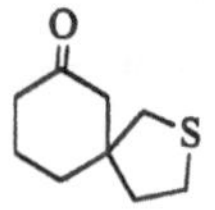

O

S

170

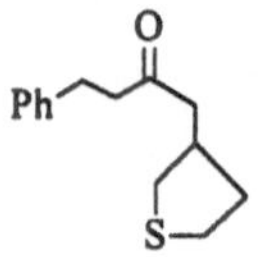

O

Ph

S

171

S   S

R R'

172a, R=Sn n-Bu₃, R'=H
b, R=R'=Me₃Sn
c, R=R'=Me₃Si

Chemical oxidations of 2-metalated 1,3-dithianes **172** using Ce(IV) or ferrocenium hexafluorophosphate results in C-M cleavage to generate the corresponding cation which reacts with a wide variety of alkenes and O-silyl enol ethers [345]. Use of the silyl compound instead of the stannyl compounds results in much poorer yields. It is suggested that this is due to more efficient cleavage of the C-Sn than the C-Si bond in the radical cation. Calculations using PM3 show this relative bond strengths in the radical cations. The cleavage of the radical cation **172a** was investigated [345] to determine if homolysis occurred to give the carbocation and tin radical or heterolysis to yield the tin cation and carbon radical (which would then be oxidized to the carbocation). Attempts to trap the carbon radical inter- or intramolecularly failed but, in the presence of CBr₄, the tin radical was trapped resulting in the formation of Bu₃SnBr in 54 % yield. Consequently, the radical cation of **172a** undergoes homolysis forming the carbocation and tin radical.

As mentioned previously in this review, α-stannylation lowers the oxidation potentials of alkyl aryl and vinyl aryl sulfides. This lowering depends on the overlap between the C-Sn bond and sulfur p-orbital. Substantial lowering of the oxidation potential on α-stannylation of 1,3-dithianes in a geometry dependent way has also been shown [412]. Thus the effect of monostannylation on the oxidation potential of 1,3-dithiane is comparable to that of alkyl or aryl groups. The angle between an equatorial C-Sn bond and sulfur p-orbital results in little reaction. However the angle between an axial C-Sn bond and sulfur p-orbital is very favorable for interaction. Consequently, the oxidation potential for **172b**, in which there is an axial C-Sn bond, is almost 1 V lower than that for 1,3-dithiane. This effect on oxidation potential is also reflected in gas phase ionization potentials. Thus the ionization potential for **172b** is ca. 1 eV lower than that for 1,3-dithiane. The effect of α-silylation on the ionization potential and oxidation potential is more modest than that of α-stannylation. The gas phase first ionization potential of thiirane is lowered by 0.59 eV by an Me$_3$Si group and 0.45 eV by a t-Bu group [413]. Similar effects are observed for a second such group and α-silylation of dimethylsulfide. The modest effect of α-silylation on lowering the oxidation potentials of phenylthioethers has been discussed above. Similarly the oxidation potential for **172c** is only modestly lower than that for alkylated 1,3-dithianes [412].

2.1.11
Sulfur Abstraction

Abstraction of a sulfur atom from a thiirane by a sulfur radical cation has been reported [352]. EPR spectroscopic studies provide insight into this process. Irradiation of thiirane with γ-rays in CF$_3$Cl at 77 K generates the corresponding radical cation. In more mobile Freons the dimeric radical cation is produced. On warming this species fragments and it was proposed that **173a** and ethene form.

173a, n=1   174a, n=1   175   176   177
 b, n=2   b, n=2
             c, n=3

A similar process has been reported [414] in the gas phase and the sulfur abstraction process can be repeated, i.e., **173a** can abstract another sulfur atom from thiirane to form **173b**. Ab initio calculations found [415] that 2c, 3e bond formation between thiirane and its radical cation is favorable but the proposed fragmentation is energetically unfavorable. Subsequent ab initio calculations have found [414] another reaction pathway leading to **174a** and ethene which is energetically favorable. This involves stepwise C-S bond cleavage in (thiirane)$_2^{\ddagger}$ to form **175** then radical displacement on S$^+$ giving **176** which then loses ethene to form **174a**. Evidence for stepwise rather than concerted loss of ethene from (thiirane)$_2^{\ddagger}$ has been found [416] in the reaction of *cis*- or *trans*-1,2-diphenyl thiirane catalyzed by Ar$_3$N‡. In this reaction a mixture of *cis*- and *trans*-**177** is formed. It is proposed that the corresponding radical cation dimer forms after oxidation of the thiirane and that it undergoes C-S cleavage to give the analogue of 175 which permits rotation about the C-C bond of the ring opened moiety isomerizing the *cis* and *trans* isomers. Furthermore, in the computational work successive sulfur atom abstractions would lead to ring expansion, i.e., **174b** rather than **173b**. This process is favorable up to the formation of the six-membered ring species **174c**. The isolation of **177** in the reaction of 1,2-diphenylthiirane catalyzed by Ar$_3$N‡ fits well with these theoretical studies although a mechanism involving intermediates **173** followed by ring closure was proposed.

2.1.12
C-C Bond Cleavage

Irradiation of the charge transfer band of complexes of 2,2,3,3-tetraarylthiiranes **178** and tetracyanoethylene results in C-C bond cleavage of the intermediary sulfur radical cation [327] to give **179**, which is an example of a distonic radical cation.

178	**179**	**180**	**181**

Such species are discussed in a section of this review. Back electron transfer from TCNE$^\mp$ to **179** yields thiocarbonyl ylide **180** which adds to TCNE to give **181** in a 1,3-dipolar cycloaddition reaction. This C-C cleavage of the radical cation of **178** is unusual because other substituted thiirane radical cations preferentially undergo C-S bond cleavage. It has also been found computationally that thiirane radical cation is more stable in the closed than open form [414, 415]. However, the four aryl substituents probably account for this unusual behavior. In addition, it has been shown [417] that there is a preference for conrotatory ring opening in this C-C cleavage reaction.

2.1.13
C-C Bond Formation

Dimer radical cation (Me$_2$S)$_2$‡ undergoes a rearrangement in which C-C bonds are formed [418]. Oxidation of Me$_2$S with ·OH at concentrations of Me$_2$S above 10 mmol l^{-1} produces EtSSEt. It is suggested that the C-C bond is formed by the following sequence. α-Deprotonation of (Me$_2$S)$_2$‡, which occurs less readily for the dimer than the monomer as mentioned above, nevertheless occurs and produces ylide **182**.

Me$_2$S $\therefore$ S$\overset{\bar{C}H_2}{\underset{Me}{}}$	Me$_2$S $\therefore$ SCH$_2$CH$_3$	
182	**183**	**184**

Rearrangement of **182** occurs with C-C bond formation affording **183**. Irradiation of $(Me_2S)_2^+$ at 400 nm in the gas phase produces an ion assigned structure **184** in which there has been C-C bond formation [419].

2.1.14
S-S Bond Formation

As already pointed out, dimer sulfur radical cations with $S \therefore S$ bonds are readily formed. This process occurs intermolecularly as well as intramolecularly. Similarly dications with S-S bonds are formed as illustrated by the formation of **139** on oxidation of **138**. The C-C bond forming reaction involving $(Me_2S)_2^+$ discussed above results in a product with an S-S bond as does the sulfur abstraction reaction delineated above. Another example of S-S bond formation which is synthetically useful involves the oxidation of dithioacetals and ketals, especially 1,3-dithiane derivatives. As already discussed, anodic oxidation of diaryl dithioacetals provides the diaryl disulfide and corresponding carbonyl compound. Similarly such studies on dialkyl dithioacetals and ketals result in S-S bond formation but polysulfides may also be formed and further oxidation of the initially formed disulfides occurs to give thiosulfinates and thiosulfonates [420]. There is also ambiguity concerning the mechanism for these reactions. In addition to the formation of the disulfide dication followed by C-S bond cleavage, as suggested for the formation of diaryl disulfides, another mechanism has been proposed. In this alternative mechanism it is proposed that the radical cation undergoes C-S cleavage and then S-S bond formation occurs [176, 178, 421]. Support for this latter mechanism is found in the cross coupling products obtained by anodic oxidation of unsymmetrical dithioacetals [420]. A further mechanistic complication is that nucleophilic attack can occur at sulfur on oxidation of 1,3-dithianes affording 1,3-dithiane-1-oxides as well as at carbon forming five-membered ring products with S-S bonds. Nevertheless, owing to the importance of dithioacetals and ketals, especially 1,3-dithiane and 1,3-dithiolane derivatives, in organic synthesis their cleavage to regenerate carbonyl compounds by electron transfer reactions has attracted considerable interest. 1,3-Dithianes and 1,3-dithiolanes are deblocked to regenerate the corresponding carbonyl compound in high yields by anodic oxidation using $(4-BrC_6H_4)_3N$ [346, 422] as a redox catalyst, methylene green [305, 306] or triphenylpyrylium [300, 301] sensitized photooxidation, or chemical oxidation with $Fe(bpy)_3(ClO_4)_3 \cdot 3H_2O$ [339] or *N*-fluoro-2,4,6-trimethylpyridinium triflate [348]. 1,3-Dithianes are also converted to their corresponding carbonyl compounds by oxidation with Ce(IV)

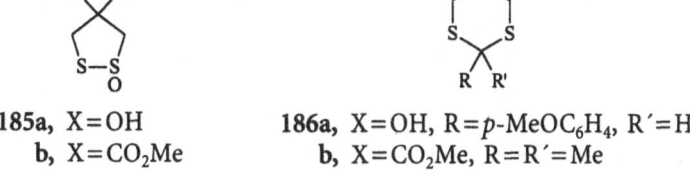

185a, X=OH
  b, X=CO_2Me

186a, X=OH, R=*p*-MeOC_6H_4, R´=H
  b, X=CO_2Me, R=R´=Me

[423, 424] or SbCl$_5$ [335]. In addition attention has been called to the sulfur products produced in these reactions and the synthesis of the natural products brugierol and isobrugierol **185a** and by anodic oxidation of **186a** [425] and methyl asparagusate **185b** by Ce(IV) oxidation of **186b** [426] have been reported.

2.2
Sulfoxide Radical Cations

here has been relatively little work on sulfoxide radical cations. γ-Radiolysis of dimethyl sulfoxide at 77 K [427] gave rise to a species with g = 2.007 and a$_H$=12 G which is tentatively assigned to Me$_2$SO$^+$. Another species with a g value of ca. 2.01 was observed and assigned to (Me$_2$SO)$_2^+$. Electrochemical oxidation of (Me$_2$N)$_2$SO in CH$_2$Cl$_2$ at 268 K gave a species with g = 2.0037, a = 13.0 G (2 N) and 12.4 (12 H) ascribed to (Me$_2$N)$_2$SO$^+$ [354]. One-electron oxidation of sulfoxides can be achieved only with oxidants with oxidation potentials greater than +2 V such as SO$_4^{-}$, CH$_3$I$^+$, (CH$_3$I)$_2^+$, or Tl^{2+} under pulse radiolysis conditions [428]. The λ_{max} for R$_2$SO$^+$ where R=Me, Et, and Pr is 300, 320, and 330 nm, respectively. The pK$_a$ values for deprotonation of the water adduct of sulfoxide radical cations (R$_2$SO $\therefore$ OH$_2$)$^+$ is estimated as 5.6, 6.1, and 6.5 for R=Me, Et, and Pr, respectively. Sulfoxide radical cations are good oxidants and readily oxidize dialkyl sulfides, dithioethers, disulfides, Br$^-$, I$^-$, and $^-$SCN. This allows the estimation of E^0 as 1.8–2.0 V vs NHE. Presumably adducts R$_2$SO $\therefore$ X are formed in the reaction of R$_2$SO$^+$ with Br$^-$, $^-$SCN, and I$^-$ but these species are not observed in the very fast redox reactions. However, Cl$^-$ reacts with R$_2$SO$^+$ to form R$_2$SO $\therefore$ Cl with λ_{max} ca. 400 nm. Oxidation of chloropromazine, triphenylamine, and zinc porphyrin by Me$_2$SO $\therefore$ Cl has been reported [429] with rate constants of 10^7–10^8 M^{-1}s^{-1}.

2.3
Distonic Sulfur Radical Cations

Distonic radical cations are species in which the charge and the radical center are separated. Interest in such species is due to calculations which suggest that such species may be more stable than the related radical cation with the charge and radical center on the same atom [430]. Ab initio calculations of $\dot{C}$H$_2\overset{+}{S}$H$_2$ indicate that it is 95 kcal/m higher in energy than CH$_3\overset{+}{S}$H using MP3/6–31G** and 86 kcal/m with zero-point vibrational correction [431] and 76 kJ/m at the MP 3/6–31G**//4–31G level [432]. Using mass spectroscopic techniques ΔH$_f^0$ for $\dot{C}$H$_2\overset{+}{S}$H$_2$ and CH$_3\overset{+}{S}$H are found to be 219 kcal/m and 212 kcal/m, respectively [433]. Ab initio calculations on $\dot{C}$H$_2\overset{+}{S}$HCH$_3$ show that it lies 82 kJ/m above Me$_2$S$^+$ with a barrier to rearrangement of 120 kJ/m at the MP 3/6–31G**//4–31G level [432]. Ab initio calculations on $\dot{C}$H$_2\overset{+}{S}$Me$_2$ show that it is 75 kJ/m higher in energy than Me$\overset{+}{S}$Et at the MP2/6–31G*//6–31G*+ZPVE level [434]. However, these distonic ions are expected to be energy minima with a significant barrier for isomerization which should permit their selective synthesis and characterization. A variety of ion-molecule reactions in the gas phase were used to synthesize $\dot{C}$H$_2\overset{+}{S}$Me$_2$ [434–438]. This species could be distinguished from Me$\overset{+}{S}$Et by the loss of CH_2 from the distonic ion but not $Me\overset{+}{S}Et$ [436]. In addition, the distonic ion abstracts MeS· from Me_2S_2 [434, 435] but $Me\overset{+}{S}Et$ only undergoes electron transfer with this species [435]. Transfer of CH_2^+· from the distonic ion to $(MeO)_3P$ is also diagnostic for this species [435]. The distonic ion is a better hydrogen atom abstractor toward thiophenol, and benzeneselenol than $Me\overset{+}{S}Et$ [435]. Indeed $Me\overset{+}{S}Et$ undergoes electron transfer predominantly with benzeneselenol rather than hydrogen atom abstraction [435]. However, it reacts 8–9 times slower with 1,4-cyclohexadiene than $Me\overset{+}{S}Et$ [435–437]. The distonic ion preferentially abstracts I· from allyl iodide whereas $Me\overset{+}{S}Et$ preferentially removes $CH_2=CHCH_2^+$ from this species [435].

2.4
Disulfide Radical Cations

Disulfide radical cations have been prepared and studied. Some EPR spectroscopic parameters for these π-radical cations [439] are listed in Table 12.

RSSR

187a, R=Me
  b, R=n-C_5H_{11}
  c, R=$CH_2CH(NH_3^+)CO_2H$
  d, R=t-Bu
  e, R=NMe_2

188a, n=0, R=R^1=H
  b, n=0, R=R^1=Me
  c, n=0, R=R^1=$(CH_2)_4$
  d, n=1, R=R^1=H
  e, n=1, R=Ph, R^1=H
  f, n=1, R=$(CH_2)_4$, CO_2H, R^1=H
  g, n=2, R=R^1=H
  h, n=3, R=R^1=H

189a, n=0, R, R^1=$(CH_2)_4$
  b, n=2, R=R^1=H
  c, n=2, R=$(CH_2)_4$
  d, n=3, R=R^1=H

190a, n=1
  b, n=2

These species were prepared by γ-radiolysis of their corresponding dialkyl disulfides **187** at 77 K, treatment of the cyclic disulfides **188–191** or their corresponding dithiols with aluminum chloride in dichloromethane or sulfuric acid, by anodic oxidation of aryl disulfides **192**, **193a**, and **194a**, by bromine oxidation of **195** and iodine oxidation of **196**.

The magnitude of the hyperfine splitting constant has a strong $\cos^2\theta$ angular dependence which renders it valuable for conformational analysis. Furthermore, variable temperature ESR spectroscopy provides a method for determining conformational barriers. These applications are well-illustrated in such

191   192   193 a, $R^1=R^2=R^3=H$ 194a, R=H
                      b, $R^1=R^2=R^3=OMe$ b, R=Me
                      c, $R^1=H$, $R^2=R^3=OMe$
                      d, $R^1=R^3=OMe$, $R^2=H$

Table 12. EPR spectroscopic parameters for disulfide radical cations

Compound	g values				a_H, G
	g_x	g_y	g_z	g_{iso}	
187a	2.032[a]		2.002[a]		9.1[a]; 9,7,7[b]; 8[c]
	2.036	2.017	2.002	2.018[b]	
	2.036	2.017	2.002	2.0183	
187b	2.035	2.018	2.003	2.019[d]	
187c	2.033	2.028	2.005	2.022[e]	
187d	2.035	2.018	2.003	2.019[f]	
	2.035	2.020	2.003	2.019[g]	
187e				2.0047[h]	7.5 (3H)[h,i]
188a				2.0193[j]	3.7[j]
188b				2.0184[k]	7.6 (1H), 1.1 (3H)[k]
188c				2.0187[k]	7.6 (1H)[k]
188d				2.0182[j], 2.0183[m]	10.0[j], 9.5[m]
188e				2.0172[k]	14.4 (1H), 11.9 (1H), 4.7 (1H), 1.2 (1H)[k]
188f				2.0183[m]	12.25 (1H), 10.6 (1H), 7.75 (1H)[m]
188g				2.0185[l]	9.5[l]
188h				2.0183[l]	9.6[l]
189a				2.0187[k]	6.3 (2H)[k]
189b				2.0185[l]	9.5[l]
189c				2.0186[k]	17.5 (2H), 4.3 (2H)[k]
189d				2.0183[l]	9.5[l]
190a				2.018[j]	6.3[j]
190b				2.0186[k]	3.8(4H)[k]
191				2.0175[k]	15.8 (2H)[k]

Compound	g values				a_H, G
	g_x	g_y	g_z	g_{iso}	
192				2.0081[p,n]	5.30, 4.44, 0.88[p,n]
				2.0079[o]	5.52 (H5), 4.56 (H3) 1.08 (H4),[o,p]
				2.0086[q]	5.25 (H5), 4.32 (H3), 0.92 (H4)[q,r]
193a				2.0112[s,q]	1.83 (H6), 0.98 (H5) 0.38 (H4), 0.1 (H3)[q,s]
194a				2.0094[t]	1.52[t]
195				2.0086[o]	1.80 (H3) 0.55 (H12) 0.43 (H11)[o,u]
196				2.0077[o]	0. 55 (H2), 0.55 (H3)[o,v]
197				2.011[w]	x

[a] [440]
[b] In D_2SO_4 [360]
[c] In $CFCl_3$ [353]
[d] [441]
[e] [442]
[f] [443]
[g] [351]
[h] [444]

[i] a_N=7.56 [444]
[j] [94]
[k] [445]
[l] [446]
[m] [447]
[n] [448]
[o] [227]
[p] a_{33S}=7.16 G. [227]

[q] [449]
[r] a_{33S}=7.33 G [449]
[s] a_{33S}=10.9 G [449]
[t] [450]
[u] a_{33S}=3.70, 4.01 G [227]
[v] a_{33S}=3.30 G [227]
[w] [451]
[x] a_N=3.15 G [451].

S—S ... S—S
195

S—S ... S—S
196

studies on cyclic disulfide radical cations [445, 447]. The EPR spectra of **187e** [444] and **197** [451] have also been reported. Interestingly, **197** is known to dimerize but both monomeric and dimeric AsF_6^- salts have been unequivocally characterized by X-ray crystallographic analysis [451]. The monomeric **197** has a disulfide bond length of 2.142–2.147 Å, which is longer than a normal S-S single bond length, suggesting localized odd electron density on each of the sulfur atoms with the SOMO antibonding across the S-S bond.

197   198

The crystal structure of dimeric **197** shows two long S···S bonds, 3.030 (1) and 2.986 (1) Å, between $S_3N_2^+$ units perhaps forming 2c, 3e bonds as shown in **198**. The EPR spectra of a variety of 1,2-dithietes **199** [94, 96, 445] and **200** [94, 445] and benzo-1,2-dithietes **201** [95, 453] have been reported. The reported g_{av} values for **199** and **200** range from 2.0144–2.0159.

199   **200**   **201**

For **201**, the g_{av} values vary from 2.0141–2.0160. However, the structure of these radical cations has been questioned as will be discussed subsequently.

Oxidation of disulfides **187a,d**, **187**, R=Et and **187**, R=i-Pr with a variety of one-electron oxidants: SO_4^-, Br_2^-, Ag^{2+}, $Ag(OH)^+$, Tl^{2+}, R_2S^+, 1,3,5-trimethoxy-benzene radical [395] or radical cations from methyl iodide [454], and **187**, R=$CH_2CH_2CO_2H$ with Br_2^- [455] in water under pulse radiolysis conditions efficiently gives the corresponding cations. These species absorb with λ_{max} 410–450 nm [455, 456]. The reaction of disulfides with $\cdot$OH is more complicated with addition to the disulfide competing equally with electron transfer [456, 457]. Oxidation of the amino disulfides obtained from cysteamine, cysteine, and penicillamine did not give the corresponding disulfide radical cation but rather species with λ_{max} near 380 nm which are tentatively assigned as RSS$\cdot$ radicals [455]. The reversible oxidation potentials could be determined by pulse radiolysis methods for dimethyl disulfide **187a**, α-lipoic acid **188f** and its carboxylate salt [458]. For **187a**, E^0 is 1.391±0.003 V for **188f** carboxylic acid E^0 is 1.13±0.01 V, and for **188f** carboxylate salt E^0 is 1.10±0.01 V all vs NHE [458]. The E^0 for **188f** is over 350 mV lower than that for **187a**. This is consistent with the lower first ionization potential of **188f** than of **187a**: 8.02 [459] and 8.97 eV [460], respectively. This difference in ionization potentials is ascribed to the S-S dihedral angle. In **188f** this angle is constrained by the five-membered ring and is 35° in the solid state as determined by X-ray crystallography [461]. In **187a** this angle is 84.7° as determined by microwave spectroscopy [462] and 83.9° as determined by an electron diffraction study [463]. The basis for the geometric effect on ionization potential is that the non-bonding p-type orbitals on each of the sulfur atoms of the disulfide have minimal interaction at a dihedral angle of 90° but maximal interaction at 0° [464–466]. As the localized orbitals interact the combined orbitals split in energy. The greater the interaction the greater the splitting. This interaction is depicted in Scheme 8. Thus in **187a** there is little interaction as indicated by ΔIE, the difference in ionization energy between the two ionizations of lowest energy, which is 0.24 eV [460]. However ΔIE, for **188f** is 1.80 eV [459] resulting in a lower first ionization potential for **188f** than **187a** (a better comparison for **188f** is **187b** whose first ionization potential is 8.70 eV and ΔIE is 0.22 eV) [460].

Table 13. Anodic oxidation potentials for disulfides

Compound	Oxidation potential
187, R=Me	1.88[a], 1.25[b], 1.08[c]
Et	1.86[a], 1.19[b]
n-Pr	1.80[a]
i-Pr	1.83[a]
n-Bu	2.01[a], 1.18[b]
s-Bu	1.96[a], 1.1[b]
t-Bu	1.19[b], 1.39
$PhCH_2$	1.38[b]
Ph	1.43[b] 1.33[c], 1.45[e], 1.75[f], 1.70[g], 1.90[h], 1.70[i]
$2\text{-MeC}_6\text{H}_4$	1.65[i]
$3\text{-MeC}_6\text{H}_4$	1.65[i]
$4\text{-MeC}_6\text{H}_4$	1.35[d]
$4\text{-}t\text{-BuC}_6\text{H}_4$	1.68[f]
$4\text{-MeOC}_6\text{H}_4$	1.54[f]
$2,5\text{-Me}_2\text{C}_6\text{H}_3$	1.60[i]
$2,6\text{-Me}_2\text{C}_6\text{H}_3$	1.70[i]
$3,5\text{-Me}_2\text{C}_6\text{H}_3$	1.65[j], 1.55[i]
$2,4\text{-(MeO)}_2\text{C}_6\text{H}_3$	1.31[f]
192	0.95[k], 0.94, 0.86[m]
193a	1.47[k]
193b	0.81, 0.94, 2.00[n]
193c	0.75, 0.84[o]
193d	0.76, 1.06[o]
194a	0.55, 0.93[p]
194b	0.475, 0.953[q]
195	0.41, 0.85[r]
196	0.24, 0.67[r]
202	0.150[s]
203	0.050[s]
204	0.68[l]
205	1.15[t]
206	0.82, 1.15, 1.65[n]
207a	1.15, 1.25, 1.64[n]
207b	0.64[o]
208	0.80[o]

[a] In CH_3CN, E_p V vs Ag/AgI [467].
[b] In CH_3CN, E_p V vs SCE [468].
[c] In CH_3CN, E_p V vs Ag/0.01 mol l^{-1} Ag$^+$ [469].
[d] In CH_3CN, E_p V vs SCE [470].
[e] In CH_3CN, E_p V vs SCE [471].
[f] In CH_3CN:CH_2Cl_2 (1:1), E_p V vs SHE [472].
[g] In CH_3CN containing CF_3CO_2H, E_p V vs Ag/AgCl [473].
[h] In CH_2Cl_2, E_p V vs Ag/AgCl [473].
[i] In CH_2Cl_2, containing CF_3CO_2H, E_p V vs Ag/AgCl [473].
[j] In CH_3CN, E_p V vs Ag/AgCl [474].
[k] In CH_3CN, $E_{1/2}$ V vs SCE [448].
[l] In CH_2Cl_2, $E_{1/2}$ V vs Ag/AgCl [475].
[m] In CH_3CN, $E_{1/2}$ V vs SCE [476].
[n] In CH_3CN, E_p V vs Ag/0.01 mol l^{-1} Ag$^+$ in CH_3CN [477].
[o] In CH_3CN, E_p V vs Ag/AgNO$_3$ in CH_3CN [478].
[p] In CH_3CN, E^0 V vs SCE [450].
[q] In PhCN, $E_{1/2}$ V vs Ag/AgCl [479].
[r] In CH_3CN, E_p V vs SCE [480].
[s] In $CHCl_3$, E_p V vs SCE [481].
[t] In PhCN, $E_{1/2}$ V vs Ag/AgCl [482].

Scheme 8

The irreversible oxidation potentials, determined by electrochemical methods, for dialkyl disulfides **187** in various solvents are listed in Table 13 [467–482]. The oxidation potentials for cyclic disulfides **192–194**, **202–208** are also listed. Compounds **202** [481], **203** [481], **207a** [477], **207b** [478] and **208** [478] undergo irreversible oxidation. Cyclic voltammetric studies of **193c** [475], **194a** [450] and **194b** [476] show two reversible one-electron oxidations. Compounds **202** [472] and **203** [479] show only one oxidation wave which is reversible in cyclic voltammetric studies. Compound **193b** [474] shows two reversible peaks and the two one-electron oxidation peaks for **193d** [475] are quasireversible. The first but not the second oxidation peak is reversible for **204** [474].

Several reactions have been reported for disulfide radical cations generated in water by pulse radiolysis techniques. These species are oxidants and undergo one-electron transfer with $Fe(CN)_6^{4-}$ at near diffusion controlled rates (ca. 10^{10} $M^{-1}s^{-1}$) [456, 457], with Fe^{2+} with rate constants [456] the order of 10^6 $M^{-1}s^{-1}$ with I^- [458] and with RS^- [484]. Disulfide radical cations, like other sulfur radical cations, do not react with O_2 [399]. On the basis of the second order decomposition of these radical cations in acidic and neutral aqueous solutions, their disproportionation to the corresponding dication as shown in Eq. (55) has been suggested [456]:

$$2(RSSR)^{+\cdot} \rightleftharpoons (RSSR)^{2+} + RSSR \tag{55}$$

Di-*tert*-butyl disulfide, **187d**, undergoes an irreversible one-electron oxidation at 1.3 V vs SCE. Controlled potential electrolysis of this material at 1.3 V in acetonitrile gives a mixture of di-*tert*-butyl tetrasulfide and *N*-*tert*-butylacetamide in high yields [470]. This suggests that the corresponding radical cation undergoes C-S bond cleavage as shown in Eq. (56):

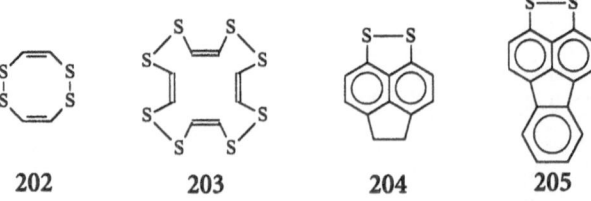

202   **203**   **204**   **205**

206   **207a,** $R^1 = R^2 = OMe$ **208**
               b, $R^1 = OMe$, $R^2 = H$

$$187d \xrightarrow{-e^-} (187d)^{\ddagger} \longrightarrow t\text{-BuSS}\cdot + t\text{-Bu}^+ \tag{56}$$

The disulfide radical then dimerizes to form di-*tert*-butyl tetrasulfide and the *tert*-butyl cation reacts with acetonitrile to form *N-tert*-butylacetamide. This C-S cleavage is undoubtedly facilitated by the formation of the relatively stable tertiary *tert*-butyl cation.

Interestingly, treatment of *p*-substituted diphenyl disulfides with $AlCl_3$ in CH_2Cl_2 results in the formation of the corresponding TH radical cations as deduced by EPR spectroscopic analysis [453]. The corresponding disulfide radical cations are not observed either because they undergo rearrangement under these conditions or because rearrangement precedes ionization. The mechanism for this transformation is not known.

Anodic oxidation of diaryl disulfides has been studied by several groups. Thiosulfonates and thiosulfinates have been reported as products in the oxidation of **187**, R=Ph [177], **193a** [485], **193b** [477], and **193c** [478]. These products apparently result from nucleophilic attack by water or perchlorate of the supporting electrolyte on the oxidized disulfide. In the presence of trialkylphosphites, the disulfide is not oxidized but rather catalyzes the oxidation of the phosphorus species to the corresponding phosphate [190]. However, the mechanisms for these reactions are not known. Anodic oxidation of disulfides under anhydrous conditions generates RS^+ or its synthetic equivalent such as $RSS^+(R)SR$. This species can then effect electrophilic reactions on aromatic rings, alkenes or alkynes. It is not clear if the disulfide radical cation cleaves the S-S bond to generate RS^+ and $RS\cdot$ and the $RS\cdot$ is then oxidized to RS^+ or if the disulfide dication is formed before S-S bond cleavage [469, 471]. Further studies are needed to clarify these mechanistic issues but it is clear that oxidation of disulfides provides a synthetically valuable method for generating RS^+ or its synthetic equivalent.

Anodic oxidation of diaryl disulfides results in polymerization [473, 474]. Thus oxidation of diphenyl disulfide **187**, R=Ph, in the presence of trifluoroacetic acid gave poly(*p*-phenylene sulfide) **209a**.

Similarly bis (3,5-dimethyl phenyl) disulfides yielded the corresponding polymer **209b** in the absence of acid. In both cases it is suggested that $ArSS^+(Ar)SAr$, formed in the reaction, effects electrophilic aromatic substitution by acting as an ArS^+ donor. This same polymerization process can also be achieved by photooxidation of bis (3,5-dimethylphenyl) disulfide in the presence of 2,3-dicyanonaphthalene as sensitizer in acetonitrile containing trifluoroacetic acid [486].

Sulfenyl ions, RS^+, or their synthetic equivalent generated by anodic oxidation of diaryl disulfides have been used to effect acetamidosulfenation of alkenes

209a, R = H   **210a**, R = Ph   **211**   **212**
   b, R = Me   **b**, R = Me

and arylthioetherification and arylthiolactonization of unsaturated alcohols and ethers. Controlled potential electrolysis of diphenyl disulfide or dimethyl disulfide at 1.3 and 1.2 V, respectively, in the presence of an alkene, such as cyclohexene, in acetonitrile produced the corresponding acetamidosulfenylation product, **210a** and **210b**, respectively [487]. The highly stereoselective formation of the *trans* products argues for the intermediacy of an episulfonium salt **211** which undergoes backside attack by acetonitrile to give **210**. It is noteworthy that dimethyl disulfide undergoes S-S cleavage in this reaction in contrast to the C-S cleavage observed on anodic oxidation of di-*tert*-butyl disulfide as outlined above. Acetamidosulfenation of steroidal alkenes by anodic oxidation of diphenyl, di-*p*-tolyl, dibenzyl, and dipropyl disulfides has also been reported [488]. Intramolecular trapping of the *S*-aryl episulfonium salt produced by anodic oxidation of diaryl disulfides provides a convenient route to arylthiolated five- and six-membered ring ethers and lactones **212** [472]. In the presence of terminal alkynes, $RC \equiv CH$, and a nucleophile, anodic oxidation of diaryl disulfides affords α-oxothioesters RCOCOSAr [471].

2.5
Polysulfide Radical Cations

A number of stable, planar, five-membered ring, 7π trisulfide radical cations have been reported recently. Reaction of hexafluoro-2-butyne with a 1:1 mixture of $S_4(AsF_6)_2$ and $S_8(AsF_6)_2$ in SO_2 provided **213a** [489]. The structure of **213a** was unequivocally established by X-ray crystallographic structural analysis. This analysis showed **213a** to be monomeric and planar in the solid state.

213a, R = CF$_3$ **214**   **215**
   b, R = CO$_2$Me

The EPR spectroscopic parameters for **213a** and **213b** are given in Table 14 and are the same as those reported previously for the corresponding 1,2-dithiete radical cations **199**, R=R^1=CF$_3$ and **199**, R=R^1=CO$_2$Me. Furthermore, in the EPR spectrum for **213b** in SO_2 two sets of ^{33}S satellites with hyperfine splitting of 8.0 and 8.9 G in a 2:1 intensity ratio could be resolved. This establishes **213b**, with two sets of sulfur atoms in a 2:1 ratio, as the correct structure in solution. Consequently, it has been suggested [489] that the previously reported 1,2-dithiete

Table 14. EPR spectroscopic parameters for trithiolium radical cations 213–219

Compound	g_x	g_y	g_z	g_{av}	hfsc, G
213a[a]				2.014	a_{19F}=1.3, a_{33S}=8.6
213b[a]				2.017	a_{33S}=8.02 (2), 8.9 (1)
214[b]				2.016	a_{33S}=8.3
216a[c]				2.012	a_H=1.06 (2H)
216b[c]				2.012	a_H=0.81 (2H)
217[d]	2.02605	2.02176	2.00355	2.01565	
218[d]	2.02454	2.01579	2.00303	2.01445	
219[e]				2.011	

[a] [489] [b] [490] [c] [492] [d] [493] [e] [494].

radical cations, whose structures were assigned on the basis of their EPR spectra, may in fact be 1,2,3-trithiolium radical cations **213**. Radical cation **214** was synthesized in an analogous way to that reported for **213a** except that CF_3CN was used instead of hexafluoro-2-butyne [490]. Its EPR spectroscopic parameters in SO_2 are given in Table 14. The SOMO is suggested to be primarily on the trisulfide moiety because of the lack of ^{14}N and ^{19}F hyperfine splitting and on the basis of ab initio MO calculations. The X-ray crystallographic structure shows this species to be monomeric and planar in the solid state. The bis-radical cation **215** has been prepared in the same way as **213a** but substituting dicyanogen for hexafluoro-2-butyne [491]. Surprisingly this species has two unpaired spins as determined by molecular susceptibility measurements and is, thereby, reminiscent of O_2. Its crystal structure has been determined by X-ray methods and consists of planar, monomeric dications. Trithiolium radical cations **216–219** annulated to benzene or phenanthrene rings have been prepared by one-electron oxidation of the corresponding trithioles.

216a, R = *i*-Pr   217, R = SMe   218   219
b, R = MeO

The oxidation potentials determined by cyclic voltammetry for formation of these radical cations from the corresponding parent trithiole are given in Table 15. In addition, **216–219** could also be prepared using chemical oxidants. Thus, the trithioles corresponding to **216a** [492], and **216b** [492] and **219** [494] on treatment with one equivalent of $NO^+PF_6^-$ afforded these radical cations. Treatment of the trithioles corresponding to **217** and **218** with $SbCl_5$ in anhydrous 1,2-dichloroethane resulted in the formation of these radical cations [492]. These radical cations underwent electron-transfer with TTF to regenerate the parent trithioles and TTF^+. Reaction of the trithiole corresponding to **216b** with

Table 15. Anodic peak potentials for obtaining radical cations **216**, **217**, and **219** from the corresponding trithioles

Radical cation formed	Oxidation potential
216a	0.73[a]
216b	0.70[a]
217	1.13[b]
219	0.69[c]

[a] In CH_3CN, $E_{1/2}$ V vs Ag/0.01 mol l^{-1} $AgNO_3$ in CH_3CN [491].
[b] $E_{1/2}$ V vs SCE [493].
[c] In CH_3CN, E_p V vs Ag/0.01 mol l^{-1} $AgNO_3$ in CH_3CN [494].

concentrated H_2SO_4 also produced this radical cation [492], which on hydrolysis gave the corresponding sulfoxides, as an approximately 2:1 mixture of 1-oxide to 2-oxide isomers, and parent trithiole as a 1:1 mixture.

References

1. (a) Stenhouse J (1868) Proc R Soc London 17:62; (b) Stenhouse J (1869) Justus Liebigs Ann Chem 149:247; (c) Graebe C (1876) Justus Liebigs Ann Chem 179:178
2. Shine HJ, Piette L (1962) J Am Chem Soc 84:4798
3. (a) Kinoshita M, Akamatu H (1962) Bull Chem Soc Jpn 35:1040; (b) Kinoshita M (1962) Bull Chem Soc Jpn 35:1137
4. Lucken EAC (1962) J Chem Soc 4963
5. Shine HJ (1994) Sulfur Rep 15:381
6. Shine HJ, Dais CF, Small RJ (1964) J Org Chem 29:21
7. (a) Sullivan PD (1968) J Am Chem Soc 90:3618; (b) Shine HJ, Sullivan PD (1968) J Phys Chem 72:1390
8. Larson SB, Simonsen SH, Martin GE, Smith K, Puig-Torres S (1984) Acta Crystallogr C40:103
9. Bock H, Rauschenbach A, Näther C, Havlas Z, Gavezzotti A, Filippini G (1995) Angew Chem Int Ed Engl 34:76
10. Joule JA (1990) Adv Heterocycl Chem 48:301
11. Simonsen SH, Lynch VM, Sutherland RG, Piórko A (1985) J Organomet Chem 290:387
12. Larson SB, Simonsen SH, Lam WW, Martin GE, Lindsay SM, Smith K (1985) Acta Crystallogr C41:1784
13. Lam WW, Martin GE, Lynch VM, Simonsen SH, Lindsay CM, Smith K (1986) J Heterocycl Chem 23:785
14. Abboud KA, Lynch VM, Simonsen SH, Piórko A, Sutherland RG (1990) Acta Crystallogr C 46:1018
15. Christie S, Piórko A, Zaworotho MJ (1994) Acta Crystallogr C50:1868
16. Martí C, Irurre J, Alvarez-Larena A, Piniella JF, Brillas E, Fajarí L, Alemán C, Juliá L (1994) J Org Chem 59:6200
17. Lovell JM, Beddoes RL, Joule JA (1996) Tetrahedron 52:4745
18. Schaefer T, Sebastian R, Beaulieu C (1991) Can J Chem 69:927
19. Amato ME, Grassi A, Irgolic KJ, Pappalardo GC, Radics L (1993) Organomet 12:775
20. Bock H, Rauschenbach A, Näther C, Kleine M, Havlas Z (1994) Chem Ber 127:2043
21. Lynch VM, Simonsen SH, Davis BE, Martin GE, Musmar MJ, Lam WW, Smith K (1994) Acta Crystallogr C50:1470

22. Martínez ES, Calleja RD, Behrens J, Berges P, Kudnig J, Wölki N, Klar G (1991) J Chem Res(S) 246
23. Hünig S, Sinzger SK, Bau R, Metzenthin T, Salbeck J (1993) Chem Ber 126:465
24. Hinrichs W, Berges P, Klar G (1987) Z Naturforsch 42b:169
25. Boduszek B, Shine HJ, Venkatachalam TK (1989) J Org Chem 54:1616
26. Sugiyama K, Shine HJ (1983) J Org Chem 48:143
27. Lochynski S, Shine HJ, Soroka M, Venkatachalam TK (1990) J Org Chem 55:2702
28. Lochynski S, Boduszek B, Shine HJ (1991) J Org Chem 56:914
29. Kochi JK (1988) Angew Chem Int Ed Engl 27:1227
30. Hoque AKMM, Shine HJ, Venkatachalam TK (1993) Coll Czech Chem Commun 58:82
31. Hammerich O, Parker VD (1984) Adv Phys Org Chem 20:55
32. Gans P, Buisson G, Duée E, Marchon J-C, Erler BS, Scholz WF, Reed CW (1986) J Am Chem Soc 108:1223
33. Balch AL, Cornman CR, Latos-Grazyński L, Olmstead MM (1990) J Am Chem Soc 112:7552
34. Halfen JA, Jazdzewski BA, Mahapatra S, Berreau LM, Wilkinson EC, Que L Jr, Tolman WB (1997) J Am Chem Soc 119:8217
35. Schmittel M, Röck M (1992) Chem Ber 125:1611
36. Schmittel M, Gescheidt G, Röck M (1994) Angew Chem Int Ed Engl 33:1961
37. Schmittel M, Langels A (1997) Angew Chem Int Ed Engl 36:392
38. Kim K, Shine HJ (1974) Tetrahedron Lett 4413
39. Kim K, Mani SR, Shine HJ (1975) J Org Chem 40:3857
40. Schulz M, Kluge R, Michaelis J (1994) Synlett 669
41. Newcomb M, Burchill MT, Deeb TM (1988) J Am Chem Soc 110:6528
42. Eberson L, Olofsson B, Svensson J-O (1992) Acta Chem Scand 46:1005
43. (a) Lu J, Yang L, Chen P, Liu Y, Liu Z (1995) Bopuxue Zazhi 12:1; (b) Lu J, Yang L, Chen P, Liu Y, Liu Z (1995) Chem Abstr 122:132,500k
44. Liu Z-L, Lu J-M, Chen P, Wang X-L, Wen X-L, Yang L, Liu Y-C (1992) J Chem Soc Chem Commun 76
45. Wen X-L, Liu Z-L, Lu J-M, Liu Y-C (1992) J Chem Soc, Faraday Soc 88:3323
46. Bancroft EE, Pemberton JE, Blount HN (1980) J Phys Chem 84:2557
47. Hammerich O, Parker VD (1972) J Electroanal Chem Interfacial Electrochem 36:App 11
48. Hammerich O, Parker VD (1973) Electrochim Acta 18:537
49. Bock H, Rauschenbach A, Ruppert K, Havlas Z (1991) Angew Chem Int Ed Engl 30:714
50. Richards TC, Bard AJ (1995) Anal Chem 67:3140
51. (a) Gupta N, Santhanam KSV (1993) Curr Sci 65:75; (b) Gupta N, Santhanam KSV (1993) Chem Abstr 119:213,181x
52. (a) Gupta N, Santhanam KSV (1995) Chem Phys 185:113; (b) Gupta N, Santhanam KSV (1994) Chem Abstr 121:267,431p
53. Bae DH, Engel PS, Hoque AKMM, Keys DE, Lee W-K, Shaw RW, Shine HJ (1985) J Am Chem Soc 107:2561
54. Engel PS, Robertson DM, Scholz J, Shine HJ (1992) J Org Chem 57:6178
55. Chen T, Shine HJ (1996) J Org Chem 61:4716
56. Chiou S, Hoque AKMM, Shine HJ (1990) J Org Chem 55:3227
57. Hoque AKMM, Lee WK, Shine HJ, Zhao D-C (1991) J Org Chem 56:1332
58. Shine HJ, Soroka M (1988) Adv Chem Ser 217:127
59. Murata Y, Shine HJ (1969) J Org Chem 34:3368
60. (a) Woo H-G, Kim S-Y, Hong L-Y, Kang H-G, Ham H-S, Kim W-G (1995) J Korean Chem Soc 39:680; (b) Woo H-G, Kim S-Y, Hong L-Y, Kang H-G, Ham H-S, Kim W-G (1995) Chem Abstr 123:199,505e
61. (a) Woo HG, Song S-J, Yang S-Y, Kim W-G (1996) J Korean Chem Soc 40:585; (b) Woo HG, Song S-J, Yang S-Y, Kim W-G (1996) Chem Abstr 125:196,477c
62. de Sorgo M, Wasserman B, Szwarc M (1972) J Phys Chem 76:3468
63. Parker VD (1984) Acc Chem Res 17:243
64. Eberson L, Blum Z, Helgée B, Nyberg K (1978) Tetrahedron 34:731

65. Pross A (1986) J Am Chem Soc 108:3537
66. Parker VD, Tilset M (1987) J Am Chem Soc 109:2521
67. Reitstöen B, Norrsell F, Parker VD (1989) J Am Chem Soc 111:8463
68. Jones G II, Huang B (1993) Tetrahedron 34:269
69. Jones G II, Huang B, Griffin SF (1993) J Org Chem 58:2035
70. Shaik SS, Pross A (1989) J Am Chem Soc 111:4306
71. Jones G II, Huang B (1992) J Phys Chem 96:9603
72. Shine HJ, Yueh W (1994) J Org Chem 59:3553
73. Zhao W, Shine HJ (1996) Tetrahedron Lett 37:1749
74. Tidwell TT (1990) Org React 39:297
75. Tidwell TT (1990) Synthesis 857
76. Pfitzer KE, Moffatt JG (1965) J Am Chem Soc 87:5661
77. Torssell K (1967) Acta Chem Scand 27:1
78. Moffatt JG (1971) J Org Chem 36:1909
79. Han DS, Shine HJ (1996) J Org Chem 61:3977
80. Shin S-R, Shine HJ (1992) J Org Chem 57:2706
81. Bosch E, Kochi JK (1995) J Chem Soc, Perkin Trans 1 1057
82. Bosch E, Kochi JK (1995) J Org Chem 60:3172
83. Shine HJ, Silber JJ, Bussey RJ, Okuyama T (1972) J Org Chem 37:2691
84. Pemberton JE, McIntire GL, Blount HN, Evans JF (1979) J Phys Chem 83:2696
85. Boduszek B, Shine HJ (1988) J Org Chem 53:5142
86. Bard AJ, Ledwith A, Shine HJ (1976) Adv Phys Org Chem 13:155
87. Shine HJ (1978) ACS Symp Ser 69:359
88. Shine HJ (1981) In: Sterling CJM, Patai S (eds) The chemistry of the sulphonium group. Wiley, Chichester, chap 14
89. Eberson L (1987) Electron transfer reactions in organic chemistry. Springer, Berlin Heidelberg New York
90. Eberson L, Shaik SS (1990) J Am Chem Soc 112:4484
91. Cho JK, Shaik SS (1991) J Am Chem Soc 113:9890
92. Bryce MR, Chesney A, Lay AK, Batsanor AS, Howard JAK (1996) J Chem Soc, Perkin Trans 1 2451
93. Lucken EAC (1963) Theor Chim Acta 1:397
94. Russell GA, Law WC, Zaleta M (1985) J Am Chem Soc 107:4175
95. Bock H, Hierholzer B, Rittmeyer P (1989) Z Naturforsch 44B:187
96. Bock H, Rittmeyer P, Stein U (1986) Chem Ber 119:3766
97. Galasso V (1976) Mol Phys 31:57
98. Howell PA, Curtis RM, Lipscomb WN (1954) Acta Crystallogr 7:493
99. Kobayashi K, Gajurel CL (1986) Sulfur Rep 7:123
100. Varma KS, Sasaki N, Clark RA, Underhill AE, Simonsen O, Becher J, Bøwardt S (1988) J Heterocycl Chem 25:783
101. Yamaguchi Y, Ueda I (1984) Acta Crystallogr C40:113
102. Bock H, Roth B, Lakshmikantham MV, Cava MP (1984) Phosphorus Sulfur 21:67
103. Schroth W, Borsdorf R, Herzschuh R, Seidler J (1970) Z Chem 10:147
104. Tani H, Kamada Y, Azuma N, Ono N (1994) Tetrahedron Lett 35:7051
105. Anderson ML, Nielsen MF, Hammerich O (1997) Acta Chem Scand 51:94
106. Mizuno M, Cava MP, Garito AF (1976) J Org Chem 41:1484
107. Sugimoto T, Sugimoto I, Kawashima A, Yamato Y, Misaki Y, Yoshida Z (1987) Heterocycles 25:83
108. Nabeshima T, Iwata S, Furukawa N, Morihashi K, Kikuchi O (1988) Chem Lett 1325
109. Magno F, Bontempelli G (1972) J Electroanal Chem Interfac Electrochem 36:389
110. Baciocchi E, Intini D, Piermattei A, Rol C, Ruzziconi R (1989) Gazz Chim Ital 119: 649
111. Elinson MN, Simonet J, Toupet L (1993) J Electroanal Chem 350:117
112. Hoffelner H, Yorgiyadi S, Wendt H (1975) J Electroanal Chem Interfac Electrochem 66:138

113. (a) Voronkov MG, Deryagina EN, Deriglazov NM, Shagun LG, Anisimova MI (1980) Zh Obsch Khim 50:2603; (b) Voronkov MG, Deryagina EN, Deriglazov NM, Shagun LG, Anisimova MI (1981) Chem Abstr 94:111,312p
114. Fox MA, Abdel-Wahab AA (1990) Tetrahedron Lett 31:4533
115. Engman L, Lind J, Merényi G (1994) J Phys Chem 98:3174; but see Merényi G, Lind J, Engman L (1996) J Phys Chem 100:8875 footnote 22
116. Sumiyoshi T, Sakai H, Kawasaki M, Katayama M (1992) Chem Lett 617
117. Tilley M, Pappas B, Pappas SP, Yagci Y, Schnabel W, Thomas JK (1989) J Imag Sci 33:62
118. Uneyama K, Torii S (1972) J Org Chem 37:367
119. Wendt H (1982) Angew Chem Int Ed Engl 21:256
120. Sullivan PD, Shine HJ (1971) J Phys Chem 75:411
121. Sullivan PD, Norman LJ (1976) J Magn Reson 23:395
122. Brunton G, Gilbert BC, Mawby RJ (1976) J Chem Soc, Perkin Trans 2 1267
123. Musso GF, Piaggio P, Dellepiane G, Borghesi A (1991) J Mol Struct (Theochem) 231:195
124. Knapczyk JW, McEwen WE (1970) J Org Chem 35:2539
125. Crivello JV (1984) Adv Polym Sci 62:1
126. Dektar JL, Hacker NP (1987) J Chem Soc Chem Commun 1591
127. Dektar JL, Hacker NP (1990) J Am Chem Soc 112:6004
128. Dektar JL, Hacker NP (1988) J Org Chem 53:1833
129. Iu K-K, Kuczynski J, Fuerniss SJ, Thomas JK (1992) J Am Chem Soc 114:4871
130. Yagci Y, Schabel W, Wilpert A, Bendig J (1994) J Chem Soc, Faraday Trans 90:287
131. Bhalero UT, Sridhar M (1994) Tetrahedron Lett 35:1413
132. Ando W, Kabe Y, Miyazaki H (1980) Photochem Photobiol 31:191
133. Foote CS, Peters JW (1971) J Am Chem Soc 93:3795
134. Liang J-J, Gu C-L, Kacher ML, Foote CS (1983) J Am Chem Soc 105:4717
135. Clennan EL, Dobrowolski P, Greer A (1995) J Am Chem Soc 117:9800
136. Clennan EL (1996) Sulfur Rep 19:171
137. Zweig A, Hodgson WG, Jura WH, Maricle DL (1963) Tetrahedron Lett 26:1821
138. Forbes W, Sullivan PD (1968) Can J Chem 46:317
139. Sullivan PD (1972) Int J Sulfur Chem A 2:149
140. Alberti A, Pedulli GF, Tiecco M, Testaferri L, Tingoli M (1984) J Chem Soc, Perkin Trans 2 975
141. Dormann E, Nowak MJ, Williams KA, Angus RO, Wudl F (1987) J Am Chem Soc 109:2594
142. Clark T (1988) J Am Chem Soc 110:1672
143. Gill PMW, Radom L (1988) J Am Chem Soc 111:4931
144. Baird NC (1977) J Chem Ed 54:291
145. Ioele M, Steenken S, Baciocchi E (1997) J Phys Chem A 101:2979
146. Boden N, Borner R, Bushby RJ, Clements J (1991) Tetrahedron Lett 32:6195
147. Bentrude WG, Martin JC (1962) J Am Chem Soc 84:1561
148. Martin JC (1978) ACS Symp Ser 69:71
149. Perkins CW, Martin JC, Arduengo AJ, Lau W, Alegria A, Kochi JK (1980) J Am Chem Soc 102:7753
150. Nakanishi W, Kusuyama Y, Ikeda Y, Iwamura H (1983) Bull Chem Soc Jpn 56:3123
151. Nakanishi W, Koike S, Inoue M, Ikeda Y, Iwamura H, Imahashi Y, Kihara K, Iwai M (1977) Tetrahedron Lett 81
152. Perkins CW, Clarkson RB, Martin JC (1986) J Am Chem Soc 108:3206
153. Gilmore JR, Mellor JM (1971) Tetrahedron Lett No 43 3977
154. Engman L, Persson J, Andersson CM, Berglund M (1992) J Chem Soc, Perkin Trans 2 1309
155. Lapouyade R, Morand J-P, Chasseau D, Amiell J, Delhaes P (1986) Synth Met 16:385
156. Lapouyade R, Morand J-P (1987) J Chem Soc Chem Commun 223
157. Hellberg J, Engman L (1987) Synth Met 19:727
158. Hellberg J, Söderholm S, Noreland J, Bietsch W, von Schütz J-U (1993) Synth Met 55/57:2102
159. Hellberg J, Söderholm S, von Schütz J-U (1991) Synth Met 41/43:2557

160. Glass RS, Andruski SW, Broeker JL, Firouzabadi H, Steffen LK, Wilson GS (1989) J Am Chem Soc 111:4036
161. Pysh ES, Yang NC (1963) J Am Chem Soc 85:2124
162. Tani H, Nii K, Masumoto K, Azuma N, Ono N (1993) Chem Lett 1055
163. Heywang G, Roth S (1991) Angew Chem Int Ed Engl 30:176
164. Heywang G, Born L, Roth S (1991) Synth Met 41/43:1073
165. Li T, Giasson R (1994) J Am Chem Soc 116:9890
166. Tani H (1995) Bull Chem Soc Jpn 68:661
167. Uneyama K, Torii S (1971) Tetrahedron Lett No 4 329
168. Matsumura Y, Yamada M, Kise N, Fujiwara M (1995) Tetrahedron 51:6411
169. Baciocchi E, Rol C, Scamosci E, Sebastiani GV (1991) J Org Chem 56:5498
170. Fuchigami T (1994) Top Curr Chem 170:1
171. Mandai T, Irie H, Kawada M, Otera J (1984) Tetrahedron Lett 25:2371
172. Yoshida J, Sugawara M, Kise N (1996) Tetrahedron Lett 37:3157
173. Balavoine G, Gref A, Fischer J-C, Lubineau A (1990) Tetrahedron Lett 31:5761
174. Amatore C, Jutand A, Mallet J-M, Meyer G, Sinaÿ P (1990) J Chem Soc Chem Commun 718
175. Mallet J-M, Meyer G, Yvelin G, Jutland A, Amatore C, Sinaÿ P (1993) Carbohydr Res 244:237
176. Canfield ND, Chambers JQ (1974) Electroanal Chem Interfac Electrochem 56:459
177. Gourcy JG, Jeminet G, Simonet J (1974) J Chem Soc Chem Commun 634
178. Gourcy J, Martigny P, Simonet J, Jeminet G (1981) Tetrahedron 37:1495
179. Yoshida J, Isoe S (1987) Chem Lett 631
180. Koizumi T, Fuchigami T, Nonaka T (1987) Chem Lett 1095
181. Koizumi T, Fuchigami T, Nonaka T (1989) Bull Chem Soc Jpn 62:219
182. Yoshida J, Matsunaga S, Murata T, Isoe S (1991) Tetrahedron 47:615
183. Cooper BE, Owen WJ (1971) J Organomet Chem 29:33
184. Yoshida J, Maekawa T, Murata T, Matsunaga S, Isoe S (1990) J Am Chem Soc 112:1962
185. Yoshida J, Tsujishima H, Nakano K, Isoe S (1994) Inorg Chim Acta 220:129
186. Yoshida J, Itoh M, Isoe S (1993) J Chem Soc Chem Commun 547
187. Yoshida J, Itoh M, Morita Y, Isoe S (1994) J Chem Soc Chem Commun 549
188. Srinivasan C, Chellamani A, Rajagopal S (1985) J Org Chem 50:1201
189. Narasaka K, Okauchi T (1991) Chem Lett 515
190. Eberson L (1983) J Am Chem Soc 105:3192
191. Eberson L (1988) New J Chem 16:151
192. Baciocchi E, Fasella E, Lanzalunga O, Mattioli M (1993) Angew Chem Int Ed Engl 32:1071
193. Venimadhavan S, Amarnath K, Harvey NG, Cheng J-P, Arnett EM (1992) J Am Chem Soc 114:221
194. Dinnocenzo JP, Todd WP, Simpson TR, Gould IR (1990) J Am Chem Soc 112:2462
195. Takemoto Y, Ohra T, Koike H, Furuse S, Iwata C (1994) J Org Chem 59:4727
196. Baciocchi E, Crescenzi C, Lanzalunga O (1997) Tetrahedron 53:4469
197. Saeva FD, Breslin DT (1989) J Org Chem 54:712
198. Saeva FD, Breslin DT, Luss HR (1991) J Am Chem Soc 113:5333
199. Saeva FD, Morgan BP, Luss HR (1985) J Org Chem 50:4360
200. Ci X, Whitten DG (1989) J Am Chem Soc 111:3459
201. Gravel D, Farmer L, Ayotte C (1990) Tetrahedron Lett 31:63
202. Gravel D, Farmer L, Denis RC, Schultz E (1994) Tetrahedron Lett 35:8981
203. Novak M, Miller A, Bruice TC, Tollin G (1980) J Am Chem Soc 102:1465
204. Glass RS (1995) Xenobiotica 25:637
205. Oae S (1992) In: Oae S, Okuyama T (eds) Organosulfur chemistry: biochemical aspects. CRC Press, Boca Raton, chap 5
206. Watanabe Y, Oae S, Iyanagi T (1982) Bull Chem Soc Jpn 55:188
207. Watanabe Y, Numata T, Iyanagi T, Oae S (1981) Bull Chem Soc Jpn 54:1163
208. Oae S, Watanabe Y, Fujimori K (1982) Tetrahedron Lett 23:1189
209. Oae S, Mikami A, Matsuura T, Ogawa-Asada K, Watanabe Y, Fujimori K, Iyanagi T (1985) Biochem Biophys Res Commun 131:567

210. Baciocchi E, Bietti M, Ioele M, Lanzalunga O, Steenken S (1997) In: Minisci F (ed) Free radicals in biology and environment. Kluwer, Dordrecht, pp109–119
211. Pryor WA, Henrickson WH Jr (1983) J Am Chem Soc 105:7114
212. Baciocchi E, Lanzalunga O, Marconi F (1994) Tetrahedron Lett 35:9771
213. Kobayashi S, Nakano M, Kimura T, Schaap AP (1987) Biochemistry 26:5019
214. Doerge DR (1986) Arch Biochem Biophys 244:678
215. Pérez U, Dunford HB (1990) Biochim Biophys Acta 1038:98
216. Pérez U, Dunford HB (1990) Biochemistry 29:2757
217. Harris RZ, Newmyer SL, Ortiz de Montellano PR (1993) J Biol Chem 268:1637
218. Ozaki S, Ortiz de Montellano PR (1995) J Am Chem Soc 117:7056
219. Baciocchi E, Lanzalunga O, Malandrucco S (1996) J Am Chem Soc 118:8973
220. LeGuillanton G, Simonet J (1983) Acta Chem Scand B 37:437
221. Aplin JT, Bauld NL (1997) J Chem Soc, Perkin Trans 2 853
222. Glass RS, Guo Q, Liu Y (1997) Tetrahedron 53:12,273
223. Geske DH, Merritt MV (1969) J Am Chem Soc 91:6921
224. Chambers JQ, Canfield ND, Williams DR, Coffen DL (1970) Mol Phys 19:581
225. Coffen DL, Chambers JQ, Williams DR, Garrett PE, Canfield ND (1971) J Am Chem Soc 93:2258
226. Bock H, Brähler G, Henkel U, Schlecker R, Seebach D (1980) Chem Ber 113:289
227. Wudl F, Smith GM, Hufnagel EJ (1970) J Chem Soc Chem Commun 1453
228. Bramwell FB, Haddon RC, Wudl F, Kaplan ML, Marshall JH (1978) J Am Chem Soc 100:4612
229. Harnden RM, Moses PR, Chambers JQ (1977) J Chem Soc Chem Commun 11
230. Moses PR, Harden RM, Chambers JQ (1977) J Electroanal Chem 84:187
231. Ferraris JP, Cowan DO, Walatka V, Perlstein JH (1973) J Am Chem Soc 95:948
232. Williams JM, Schultz AJ, Geiser U, Carlson KD, Kini AM, Wang KH, Kwok W-K, Whangbo M-H, Schirber JE (1991) Science 252:1501
233. McKenzie RH (1997) Science 278:820
234. Williams JM, Ferraro JR, Thorn RJ, Carlson KD, Geiser U, Wang HH, Kini AM, Whangbo M-H (1992) Organic superconductors. Prentice Hall, Englewood Cliffs
235. Jørgensen T, Hansen TK, Becher J (1994) Chem Soc Rev 41
236. Adam M, Müllen K (1994) Adv Mater 6:439
237. Schukat G, Fanghänel E (1996) Sulfur Rep 18:1
238. Hoffmann R (1987) Angew Chem Int Ed Engl 26:846
239. Baumgarten M, Müllen K (1994) Top Curr Chem 169:1
240. Murphy JA, Rasheed F, Roome SJ, Lewis N (1996) J Chem Soc Chem Commun 737
241. Skrydstrup T (1997) Angew Chem Int Ed Engl 36:345
242. Bashir N, Callaghan O, Murphy JA, Ravishanker T, Roome SJ (1997) Tetrahedron Lett 38:6255
243. Rao DNR, Symons MCR (1983) J Chem Soc, Perkin Trans 2 135
244. Shiotani M, Nagata Y, Tasaki M, Sohma J, Shida T (1983) J Phys Chem 87:1170
245. Davies AG, Julia L, Yazdi SN (1989) J Chem Soc, Perkin Trans 2 239
246. Roncali J (1992) Chem Rev 92:711
247. Pool R (1994) Science 263:1700
248. Shi G, Jin S, Xue G, Li C (1995) Science 267:994
249. Smith JR, Cox PA, Campbell SA, Ratcliffe NM (1995) J Chem Soc, Farad Trans 91:2331
250. Tabakovic I, Maki T, Miller LL, Yu Y (1996) J Chem Soc Chem Commun 1911
251. Garnier F, Hajlaoui R, Yassar A, Srivastava P (1994) Science 265:1684
252. Horowitz G, Peng X-Z, Fichou D, Garnier F (1991) J Mol Electron 7:85
253. Evans CH, Scaiano JC (1990) J Am Chem Soc 112:2694
254. Samdal S, Samuelson EJ, Volden HV (1993) Synth Met 59:259
255. DiCésare N, Belletête M, Raymond F, Lederc M, Durocher G (1997) J Phys Chem A 101:776
256. Alemán C, Brillas E, Davies AG, Fajarí L, Giró D, Juliá L, Pérez JJ, Rius J (1993) J Org Chem 58:3091

257. Lemaire M, Büchner W, Garreau R, Hoa HA, Guy A, Roncali J (1990) J Electroanal Chem 281:293
258. Caspar JV, Ramamurthy V, Corbin DR (1991) J Am Chem Soc 113:600
259. Yu Y, Gunic E, Zinger B, Miller LL (1996) J Am Chem Soc 118:1013
260. Wintgens V, Valat P, Garnier F (1994) J Phys Chem 98:228
261. Xu Z-G, Horowitz G (1992) J Electroanal Chem 335:123
262. Engelmann G, Stößer R, Koßmehl G, Jugelt W, Welzel H-P (1996) J Chem Soc, Perkin Trans 2 2015
263. Graf DD, Duan RG, Campbell JP, Muller LL, Mann KR (1997) J Am Chem Soc 119:5888
264. Hill MG, Mann KR, Muller LL, Penneau J-F (1992) J Am Chem Soc 114:2728
265. Bäuerle P, Segelbacher U, Maier A, Mehring M (1993) J Am Chem Soc 115:10,217
266. Guay J, Diaz A, Wu R, Tour JM (1993) J Am Chem Soc 115:1869
267. Roncali J, Giffard M, Jubault M, Gorgues A (1993) J Electroanal Chem 361:185
268. Bäuerle P, Fischer T, Bidlingmeier B, Stabel A, Rabe JP (1995) Angew Chem Int Ed Engl 34:303
269. Kuroda M, Nakayama J, Hoshino M, Furusho N (1992) Tetrahedron Lett 33:7553
270. Tanaka K, Ago H, Yamabe T, Ishikawa M, Ueda T (1994) Organometallics 13:3496
271. Ago H, Kuga T, Yamabe T, Tanaka K, Kunai A, Ishikawa M (1997) Chem Mater 9:1159
272. Suzuki T, Shiohara H, Monobe M, Sakimura T, Tanaka S, Yamashita Y, Miyashi T (1992) Angew Chem Int Ed Engl 31:455
273. Roncali J, Giffard M, Jubault M, Gorgues A (1993) Synth Met 60:163
274. Eberson L, Hartshorn MP, Persson O, Radner F (1997) Acta Chem Scand 51:492
275. Butts CP, Eberson L, Hartshorn M, Radner F, Robinson WT, Wood BR (1997) Acta Chem Scand 51:839
276. Fujitsuka M, Sato T, Shimidzu T, Watanabe A, Ito O (1997) J Phys Chem A 101:1056
277. Yoneda S, Tsubouchi A, Ozaki K (1987) Nippon Kagaku Kaishi 1328
278. Tsubouchi A, Matsumura N, Inoue H (1991) J Chem Soc Chem Commun 520
279. Ishii A, Nakayama J, Kazami J-I, Ida Y, Nakamura T, Hoshino M (1991) J Org Chem 56:78
280. Symons MCR (1984) Chem Soc Rev 13:393
281. von Sonntag C, Schuchmann H-P (1980) In: Patai S (ed) The chemistry of ethers, crown ethers, hydroxyl groups, their sulphur analogues, suppl E, pt 2. Wiley, Chichester, chap 24
282. Asmus K-D (1984) In: Packer L (ed) Methods in enzymology, vol 105. Academic Press, Orlando, pp 167–178
283. von Sonntag C (1987) The chemical basis of radiation biology. Taylor Francis, London
284. Waltz WL (1988) In: Fox MA, Chanon M (eds), Photoinduced electron transfer, pt B. Elsevier, Amsterdam, pp 57–109
285. Takamuku S, Yamamoto Y (1991) In: Tabata Y (ed) Pulse radiolysis. CRC Press, Boca Raton, chap 18
286. Pienta NJ (1988) In: Fox MA Chanon M (eds) Photoinduced electron transfer, pt C. Elsevier, Amsterdam pp, 421–486
287. van de Sande CC (1980) In: Patai S (ed) The chemistry of ethers, crown ethers, hydroxyl groups, their sulphur analogues, pt 1. Wiley, Chichester, chap 7
288. Nibbering NMM, Ingemann S, de Koning LJ (1996) In: Baer T, Ng CY, Powis I (eds) The structure, energetics, dynamics of organic ions. Wiley, New York, pp 283–290
289. Merényi G, Lind J, Engman L (1996) J Phys Chem 100:8875
290. Meissner G, Henglein A, Beck G (1967) Z Naturforsch B 22:13
291. Bonifačić M, Möckel H, Bahnemann D, Asmus K-D (1975) J Chem Soc, Perkin Trans 2 675
292. Asmus K-D, Bahnemann D, Bonifačić M, Gillis HA (1977) Faraday Discuss Chem Soc 63:213
293. Bobrowski K, Schöneich C (1993) J Chem Soc Chem Commun 795
294. Mohan H, Mittal JP (1992) J Chem Soc, Perkin Trans 2 207
295. Gu M, Turecek F (1992) J Am Chem Soc 114:7146
296. Hynes AJ, Wine PH, Semmes DH (1986) J Phys Chem 90:4148
297. McKee ML (1993) J Phys Chem 97:10,971

298. Schnöneich C, Bobrowski K (1993) J Am Chem Soc 115:6538
299. Bowen JR, Yang SF (1975) Photochem Photobiol 21:201
300. Garcia H, Iborra S, Miranda MA, Primo J (1989) New J Chem 13:805
301. Kamata M, Kato Y, Hasegawa E (1991) Tetrahedron Lett 32:4349
302. Fasani E, Freccero M, Mella M, Albini A (1997) Tetrahedron 53:2219
303. Pandey B, Bal SY, Khire UR (1989) Tetrahedron Lett 30:4007
304. Mathew L, Sankararaman S (1993) J Org Chem 58:7576
305. Epling GA, Wang Q (1992) Synlett 335
306. Epling GA, Wang Q (1992) Tetrahedron Lett 33:5909
307. Davidson RS, Pratt JE (1983) Tetrahedron Lett 24:5903
308. Cohen SG, Ojanpera S (1975) J Am Chem Soc 97:5633
309. Inbar S, Linschitz H, Cohen SG (1982) J Am Chem Soc 104:1679
310. Jones G II, Malba V (1985) New J Chem 9:5
311. Jones G II, Malba V, Bergmark WR (1986) J Am Chem Soc 108:4214
312. Bobrowski K, Marciniak B, Hug GL (1994) J Photochem Photobiol A 81:159
313. Bobrowski K, Marciniak B, Hug GL (1992) J Am Chem Soc 114:10,279
314. Marciniak B, Bobrowski K, Hug GL (1993) J Phys Chem 97:11,937
315. Bobrowski K, Hug GL, Marciniak B, Kozubek H (1994) J Phys Chem 98:537
316. Marciniak B, Bobrowski K, Hug GL, Rozwadowski J (1994) J Phys Chem 98:4854
317. Marciniak B, Hug GL, Bobrowski K, Kozubek H (1995) J Phys Chem 99:13,560
318. Jensen JL, Miller BL, Zhang X, Hug GL, Schöneich C (1997) J Am Chem Soc 119:4749
319. Bobrowski K, Hug GL, Marciniak B, Miller B, Schöneich C (1997) J Am Chem Soc 119:8000
320. Hasegawa E, Brumfeld MA, Mariano PS (1988) J Org Chem 53:5435
321. Yoon U-C, Kim H-J, Mariano PS (1989) Heterocycles 29:1041
322. Yoon U-C, Kim YC, Choi JJ, Kim DU, Mariano PS, Cho I-S, Jeon YT (1992) J Org Chem 57:1422
323. Ikeno T, Harada M, Arai N, Narasaka K (1997) Chem Lett 169
324. Griesbeck AG, Mauder H, Müller I, Peters E-M, Peters K, von Schnering HG (1993) Tetrahedron Lett 34:453
325. Hu S, Neckers DC (1997) Tetrahedron 53:2751
326. Saito I, Takayama M, Sakurai T (1994) J Am Chem Soc 116:2653
327. Kamata M, Miyashi T (1989) J Chem Soc Chem Commun 557
328. Glass RS, Broeker JL, Anklam E, Asmus K-D (1992) Tetrahedron Lett 33:1721
329. Coleman BR, Glass RS, Setzer WN, Prabhu UDG, Wilson GS (1982) Adv Chem Ser 201:417
330. Wilson GS (1990) In: Chatgilialoglu C, Asmus K-D (eds), Sulfur-centered reactive intermediates in chemistry, biology. Plenum Press, New York, pp83–92
331. Elinson MN, Simonet J (1992) J Electroanal Chem 336:363
332. Frey JE, Aiello T, Beaman DN, Hutson H, Lang SR, Puckett JF (1995) J Org Chem 60:2891
333. Lichtenberger DL, Johnston RL, Hinkelmann K, Suzuki T, Wudl F (1990) J Am Chem Soc 112:3302
334. Block E, Glass RS, DeOrazio R, Lichtenberger DL, Pollard JR, Russell EE, Schroeder TB, Thiruvazhi M, Toscano PJ (1997) Synlett 525
335. Kamata M, Otogawa H, Hasegawa E (1991) Tetrahedron Lett 32:7421
336. Musker WK, Wolford TL, Roush PB (1978) J Am Chem Soc 100:6416
337. Musker WK (1980) Acc Chem Res 13:200
338. Siddique RM, Winfield JM (1989) Can J Chem 67:1780
339. Murase M, Kotani E, Tobinaga S (1986) Chem Pharm Bull 34:3595
340. Riley DP, Smith MR, Correa PE (1988) J Am Chem Soc 110:177
341. Dinnocenzo JP, Banach TE (1986) J Am Chem Soc 108:6063
342. Gilbert BC, Hodgeman DKC, Norman ROC (1973) J Chem Soc, Perkin Trans 2 1748
343. Gilbert BC, Marriott PR (1979) J Chem Soc, Perkin Trans 2 1425
344. Pryor WA, Jin X, Squadrito GL (1994) Proc Nat Acad Sci USA 91:11,173
345. Narasaka K, Arai N, Okauchi T (1993) Bull Chem Soc Jpn 66:2995
346. Platen M, Steckhan E (1984) Chem Ber 117:1679

347. Tanemura K, Dohya H, Imamura M, Suzuki T, Horaguchi T (1994) Chem Lett 965
348. Kiselyov AS, Strekowski L, Semenov VV (1993) J Heterocycl Chem 30:329
349. Kiselyov AS, Strekowski L, Semenov VV (1993) Tetrahedron 49:2151
350. (a) Wang JT, Williams F (1981) J Chem Soc Chem Commun 1184; (b) Wang JT, Williams F (1983) J Chem Soc Chem Commun 980
351. Bonazzola L, Michaut JP, Roncin J (1985) J Chem Phys 83:2727
352. Qin X-Z, Meng Q, Williams F (1987) J Am Chem Soc 109:6778
353. Rao DNR, Symons MCR, Wren BW (1984) J Chem Soc, Perkin 2 1681
354. Gara WG, Giles JRM, Roberts BP (1979) J Chem Soc, Perkin Trans 2 1444
355. Kira M, Nakazawa H, Sakurai H (1986) Chem Lett 497
356. Bonazzola L, Michaut JP, Roncin J (1988) Can J Chem 66:3050
357. Brown TG, Hirschon AS, Musker WK (1981) J Phys Chem 85:3767
358. Izuoka A, Kobayashi M (1981) Chem Lett 1603
359. Nelsen SF, Steffek DJ, Cunkle GT, Gannett PM (1982) J Am Chem Soc 104:6641
360. Petersen RL, Nelson DJ, Symons MCR (1978) J Chem Soc, Perkin Trans 2 225
361. Davies MJ, Gilbert BC, Norman ROC (1984) J Chem Soc, Perkin Trans 2 503
362. Tamaoki M, Serita M, Shiratori Y, Itoh K (1989) J Phys Chem 93:6052
363. Illies AJ, Livant P, McKee ML (1988) J Am Chem Soc 110:7980
364. Deng Y, Illies AJ, James MA, McKee ML, Peschke M (1995) J Am Chem Soc 117:420
365. James MA, McKee ML, Illies AJ (1996) J Am Chem Soc 118:7836
366. Mönig J, Goslich R, Asmus K-D (1986) Ber Bunsen-Ges Phys Chem 90:115
367. Clark T (1990) In: Chatgilialoglu C, Asmus K-D (eds) Sulfur-centered reactive intermediates in chemistry, biology. Plenum Press, New York, pp 13–18
368. Asmus K-D (1979) Acc Chem Res 12:436
369. Asmus K-D (1990) In: Chatgilialoglu C, Asmus K-D (eds) Sulfur-centered reactive intermediates in chemistry, biology. Plenum Press, New York, pp155–172
370. Chaudhri SA, Mohan H, Anklam E, Asmus K-D (1996) J Chem Soc, Perkin Trans 2 383
371. Asmus K-D, Bahnemann D, Fischer Ch-H, Veltwich D (1979)J Am Chem Soc 101:5322
372. Mahling S, Asmus K-D, Glass RS, Hojjatie M, Sabahi M, Wilson GS (1987) J Org Chem 52:3717
373. Mohan H (1990) J Chem Soc, Perkin Trans 2 1821
374. Asmus K-D, Göbl M, Hiller K-O, Mahling S, Mönig J (1985) J Chem Soc, Perkin Trans 2 641
375. Steffen LK, Glass RS, Sabahi M, Wilson GS, Schöneich C, Mahling S, Asmus K-D (1991) J Am Chem Soc 113:2141
376. Musker WK, Surdhar PS, Ahmad R, Armstrong DA (1984) Can J Chem 62:1874
377. Musker WK, Hirschon AS, Doi JT (1978) J Am Chem Soc 100:7754
378. Glass RS, Sabahi M, Singh WP (1992) J Org Chem 57:2683
379. Carmichael I (1997) Acta Chem Scand 51:567
380. Bonifačić M, Asmus K-D (1980) J Chem Soc, Perkin Trans 2 758
381. Anklam E, Mohan H, Asmus K-D (1988) J Chem Soc, Perkin Trans 2 1297
382. Hungerbühler H, Guha SN, Asmus K-D (1991) J Chem Soc Chem Commun 999
383. Tobien T, Hungerbühler H, Asmus K-D (1994) Phosphorus, Sulfur, Silicon 95/96:249
384. Colson A-O, Sevilla MD (1991) J Phys Chem 95:6487
385. Werst DW (1992) J Phys Chem 96:3640
386. Werst DW, Trifunac AD (1991) J Phys Chem 95:3466
387. Wilson GS, Swanson DD, Klug JT, Glass RS, Ryan MD, Musker WK (1979) J Am Chem Soc 101:1040
388. Ryan MD, Swanson DD, Glass RS, Wilson GS (1981) J Phys Chem 85:1069
389. Furukawa N, Kawada A, Kawai T (1984) J Chem Soc Chem Commun 1151
390. Fujihara H, Kawada A, Furukawa N (1987) J Org Chem 52:4254
391. Fujihara H, Akaishi R, Furukawa N (1987) J Chem Soc Chem Commun 930
392. Iwasaki F, Toyoda N, Akaishi R, Fujihara H, Furukawa N (1988) Bull Chem Soc Jpn 61:2563
393. Doi JT, Musker K (1978) J Am Chem Soc 100:3533

394. Fujihara H, Furukawa N (1989) J Mol Struct 186:261
395. Bonifačić M, Asmus K-D (1976) J Phys Chem 80:2426
396. Chaudhri SA, Göbl M, Freyholdt T, Asmus K-D (1984) J Am Chem Soc 106:5988
397. Bonifačić M, Weiss J, Chaudhri SA, Asmus K-D (1985) J Phys Chem 89:3910
398. Schöneich C, Aced A, Asmus K-D (1993) J Am Chem Soc 115:11,376
399. Schäfer K, Bonifačić M, Bahnemann D, Asmus K-D (1978) J Phys Chem 82:2777
400. Bonifačić M, Asmus K-D (1986) J Org Chem 51:1216
401. Kimura M, Koie K, Matsubara S, Sawaki Y, Iwamura H (1987) J Chem Soc Chem Commun 122
402. Göbl M, Asmus K-D (1984) J Chem Soc, Perkin Trans 2 691
403. Miller BL, Williams TD, Schöneich C (1996) J Am Chem Soc 118:11,014
404. Miller K-O, Masloch B, Göbl M, Asmus K-D (1981) J Am Chem Soc 103:2734
405. Bobrowski K, Schöneich C, Holcman J, Asmus K-D (1991) J Chem Soc, Perkin Trans 2 353
406. Bobrowski K, Schöneich C, Holcman J, Asmus K-D (1991) J Chem Soc, Perkin Trans 2 975
407. Platen M, Steckhan E (1984) Justus Liebigs Ann Chem 1563
408. Iranpoor N, Owji J (1990) Synth Commun 20:1047
409. Iranpoor N, Owji J (1991) Tetrahedron 47:149
410. Kamata M, Murayama K, Miyashi T (1989) Tetrahedron Lett 30:4129
411. Mandai T, Yasunaga H, Kawada M, Otera J (1984) Chem Lett 715
412. Glass RS, Radspinner AM, Singh WP (1992) J Am Chem Soc 114:4921
413. Block E, Yencha AJ, Aslam M, Eswarakrishnan V, Luo J, Sano A (1988) J Am Chem Soc 110:4748
414. Ekern S, Illies A, Peschke M (1993) J Am Chem Soc 115:12,510
415. Gill PMW, Weatherall P, Radom L (1989) J Am Chem Soc 111:2782
416. Kamata M, Murayama Y, Suzuki T, Miyashi T (1990) J Chem Soc Chem Commun 827
417. Glass RS, Jung W (1994) J Am Chem Soc 116:1137
418. Goslich R, Weiss J, Möckel HJ, Mönig J, Asmus K-D (1985) Angew Chem Int Ed Engl 24:73
419. Syage JA, Pollard JE, Cohen RB (1991) J Phys Chem 95:8560
420. Porter QN, Utley JHP, Machion PD, Pardini VL, Schumacher PR, Viertler H (1984) J Chem Soc, Perkin Trans 2 973
421. Martigny P, Simonet J (1980) Electroanal Chem 111:133
422. Schultz N, Toteberg-Kaulen S, Dapperheld S, Heyer J, Platen M, Schumacher K, Steckhan E (1987) In: Torii S (ed) Recent advances in electroorganic synthesis. Elsevier, Amsterdam, pp125–172
423. Ho TL, Ho HC, Wong CW (1972) J Chem Soc Chem Commun 791
424. Cristau H-J, Chabaud B, Labaudinière R, Christol H (1981) Synth Commun 11:423
425. Glass RS, Petsom A, Wilson GS, Martínez R, Juaristi E (1986) J Org Chem 51:4337
426. Glass RS, Singh WP, Mobashar RM, Petsom A, Wilson GS, Martínez R, Ordóñez M, Juaristi E (1995) Sulfur Lett 18:259
427. Symons MCR (1976) J Chem Soc, Perkin Trans 2 908
428. Kishore K, Asmus K-D (1989) J Chem Soc, Perkin Trans 2 2079
429. Kumar M, Neta P (1992) J Phys Chem 96:3350
430. Yates BF, Bouma WJ, Radom L (1986) Tetrahedron 42:6225
431. Yates BF, Bouma WJ, Radom L (1984) J Am Chem Soc 106:5805
432. Nobes RH, Bouma WJ, Radom L (1984) J Am Chem Soc 106:2774
433. Holmes JL, Lossing FP, Terlouw JK, Burgers PC (1983) Can J Chem 61:2305
434. Smith RL, Chyall LJ, Stirk KM, Kenttämaa HI (1993) Org Mass Spectrom 28:1623
435. Stirk KM, Orlowski JC, Leeck DT, Kenttämaa HI (1992) J Am Chem Soc 114:8604
436. van Amsterdam MW, Zappey HW, Ingemann S, Nibbering NMM (1993) Org Mass Spectrom 28:30
437. van Amsterdam MW, Staneke PO, Ingemann S, Nibbering NMM (1993) Org Mass Spectrom 28:919
438. Smith RL, Franklin RL, Stirk KM, Kenttämaa HI (1993) J Am Chem Soc 115:10,348
439. Glidewell C (1984) J Chem Soc, Perkin Trans 2 407
440. Gillbro T (1974) Chem Phys 4:476

441. Truby FK (1964) J Chem Phys 40:2768
442. Box HC, Freund HG (1964) J Chem Phys 41:2571
443. Chandra H, Rao DNR, Symons MCR (1983) J Chem Res (S) 68
444. Bock H, Schultz W, Stein U (1981) Chem Ber 114:2632
445. Russell GA, Law WC (1990) In: Chatgilialoglu C, Asmus K-D (eds) Sulfur-centered reactive intermediates in chemistry, biology. Plenum, New York, pp173–183
446. Bock H, Stein U (1980) Angew Chem Int Ed Engl 19:834
447. Bock H, Stein U, Semkow A (1980) Chem Ber 113:3208
448. Zweig A, Hoffman AK (1965) J Org Chem 30:3997
449. Bock H, Brähler G, Dauplaise D, Meinwald J (1981) Chem Ber 114:2622
450. Wudl F, Schafer DE, Miller B (1976) J Am Chem Soc 98:252
451. Roesky HW, Hamza A (1976) Angew Chem Int Ed Engl 15:226
452. Gillespie RJ, Kent JP, Sawyer JF (1981) Inorg Chem 20:3784
453. Giordan J, Bock H (1982) Chem Ber 115:2548
454. Mohan H, Asmus K-D (1988) J Phys Chem 92:118
455. Elliot AJ, McEachern RJ, Armstrong DA (1981) J Phys Chem 85:68
456. Bonifačić M, Schäfer K, Möckel H, Asmus K-D (1975) J Phys Chem 79:1496
457. Möckel H, Bonifačić M, Asmus K-D (1974) J Phys Chem 78:282
458. Bonifačić M, Asmus K-D (1986) J Chem Soc, Perkin Trans 2 1805
459. Guimon M-F, Guimon C, Pfister-Guillouzo G (1975) Tetrahedron Lett 441
460. Wagner G, Bock H (1974) Chem Ber 107:68
461. Karle IL, Estlin JA, Britts K (1967) Acta Crystallogr 22:567
462. Sutter D, Dreizler H, Rudolph HD (1965) Z Naturforsch A 20:1676
463. Beagley B, McAloon KT (1971) Trans Faraday Soc 67:3216
464. Gleiter R, Spanget-Larsen J (1979) Top Curr Chem 86:139
465. Klessinger M, Rademacher P (1979) Angew Chem Int Ed Engl 18:826
466. Brown RS, Jørgensen FS, (1984) In: Brundle CR, Baker AD (eds) Electron spectroscopy: theory, techniques, and applications, vol 5. Academic Press, London, pp 1–122
467. Machio PD, Pardini VL, Viertler H (1984) IV Simp Bras Electroquim Electroanal 289
468. Masui M, Mizuki Y, Ueda C, Ohmori H (1984) Chem Pharm Bull 32:1236
469. Bewick A, Coe DE, Libert M, Mellor JM (1983) J Electroanal Chem 144:235
470. Elothmani D, Do QT, Simonet J, Le Guillanton G (1993) J Chem Soc Chem Commun 715
471. Boryczka S, Elothmani D, Do QT, Simonet J, Le Guillanton G (1996) J Electrochem Soc 143:4027
472. Töteberg-Kaulen S, Steckhan E (1988) Tetrahedron 44:4389
473. Yamamoto K, Tsuchida E, Nishide H, Yoshida S, Park Y-S (1992) J Electrochem Soc 139:2401
474. Yamamoto K, Iwasaki N, Nishide H, Tsuchida E (1992) Eur Polym J 28:341
475. Aso Y, Yui K, Miyoshi T, Otsubo T, Ogura F, Tanaka J (1988) Bull Chem Soc Jpn 61:2013
476. Dauplaise D, Meinwald J, Scott JC, Temkin H (1978) Ann NY Acad Sci 313:382
477. Cariou M, Douadi T, Simonet J (1996) New J Chem 20:1031
478. Stender K-W, Klar G, Knittel D (1985) Z Naturforsch B 40:774
479. Otsubo T, Sukenobe N, Aso Y, Ogura F (1987) Chem Lett 315
480. Svensmark B, Hammerich O (1991) In: Lund H, Baizer MM (eds) Organic electrochemistry, 3rd edn. Dekker, New York, p 663
481. Shimizu T, Iwata K, Kamigata N (1996) Angew Chem Int Ed Engl 35:2357
482. Miyamoto H, Yui K, Aso Y, Otsubo T, Ogura F (1986) Tetrahedron Lett 27:2011
483. Bontempelli G, Magno F, Mazzocchin G-A (1973) J Electroanal Chem Interfacial Electrochem 42:57
484. Bonifačić M, Asmus K-D (1984) Int J Radiat Biol 46:35
485. Dakova B, Carbonelle P, Walcarius A, Lamberts L, Evers M (1992) Electrochim Acta 37:725
486. Yamamoto K, Oyaizu K, Tsuchida E (1993) Chem Lett 1101
487. Bewick A, Coe DE, Mellor JM, Owton WM (1985) J Chem Soc, Perkin Trans 1 1033

488. Mellor JM, de Milano DLB (1986) J Chem Soc, Perkin Trans 1 1069
489. Cameron TS, Haddon RC, Mattar SM, Parsons S, Passmore J, Ramirez AP (1992) J Chem Soc, Dalton Trans 1563
490. Cameron TS, Haddon RC, Mattar SM, Parsons S, Passmore J, Ramirez AP (1992) Inorg Chem 31:2274
491. Boyle PD, Parsons S, Passmore J, Wood DJ (1993) J Chem Soc Chem Commun 199
492. Ogawa S, Saito S, Kikuchi T, Kawai Y, Niizuma S, Sato R (1995) Chem Lett 321
493. Fanghänel E, Herrmann R, Naarmann H (1995) Tetrahedron 51:2533
494. Ogawa S, Nobuta S, Nakayama R, Kawai Y, Niizuma S, Sato R (1996) Chem Lett 757

New Aspects of Hypervalent Organosulfur Compounds

Naomichi Furukawa, Soichi Sato

Tsukuba Advanced Research Alliance Center, Department of Chemistry, University
of Tsukuba, Tsukuba, Ibaraki 305, Japan
E-mail: furukawa@staff.chem.tsukuba.ac.jp

The chemistry of sulfuranes has developed greatly and has attracted attention of many organic chemists over the past two decades. This work describes the historical background, the
development, and the synthesis of sulfuranes and persulfuranes. The concept of hypervalent
bonding was introduced by Rundle and Musher; before the pioneering work of Martin and
Kapovits on the preparation of sulfuranes, these compounds were considered only to be
unstable intermediates or simple transition states in nucleophilic substitution reactions at sulfur atoms. There are many reports of unstable sulfuranes being formed during the substitution reactions of dicoordinated, tricoordinated and even tetracoordinated sulfur atoms with
various nucleophiles. However, neither direct nor spectroscopic evidence for the formation of
sulfuranes had been reported until in the 1970s, when Martin found methods for syntheses of
the stable sulfuranes. Ligand exchange and coupling reactions are typical reactions of tricoordinated sulfur compounds with nucleophiles, and are considered to be diagnostic for the
formation of sulfuranes. According to Martin's definition, stable sulfuranes should contain
electronegative substituents to provide the hypervalent bonds. However, quite recently, sulfuranes bearing only carbon substituents have been synthesized. This work describes both the
intermediary formation of sulfuranes and stable isolable sulfuranes, together with their chemical behaviors.

Keywords: Hypervalency, Sulfurane, Persulfonium salt, Persulfurane, Ligand exchange reaction,
Ligand coupling reaction, Pseudorotation, Apical bond, Equatorial bond, Trigonal bipyramidal, Octahedral, Three-center four-electron bond

Topics in Current Chemistry, Vol. 205
© Springer-Verlag Berlin Heidelberg 1999

1
General Introduction to Hypervalent Compounds [1]

Lewis' "Octet Rule" has long been a central dogma governing the chemical bonding of organic molecules, which are composed normally from the second row elements, namely carbon as a central atom and nitrogen, oxygen, hydrogen, and some other elements as auxiliaries [2].

The orbital hybridization of s- and p-orbitals proposed by Pauling is also a centrally important and general concept to explain the chemical bonding of organic compounds [3]. However, in the past several decades there have been a few unusual molecules reported which do not obey the octet rule, for example, $PhSCl_3$, $PhSF_3$, SF_6, I_3^-, etc. [4]. The central atoms of these molecules are composed of elements below the third row and from the 14th to 17th group in the periodic table, although normally these elements also provide organic molecules possessing octets. The characteristic bonding modes for these molecules have been known to possess stable decet [5] or even dodecet structures [6], and they have been considered to be the few exceptions to the octet rule. Such decet and dodecet molecules have, however, been well known in inorganic compounds which use d- and f-orbitals for making chemical bonds. Therefore, these findings indicate the existence of a new field between the inorganic and organic regions which we now call heteroatom chemistry [7]. There are two ways in which the chemical bond of SF_4 can be explained. One is the new hybridization using one $3d$-orbital of the sulfur atom with normal $3sp^3$ resulting in the formation of five coordinate bonds $3sp^3d$, fitting the requirement for a decet [3]. The other is resonance hybridization among the conceivable ionic forms, e.g., in the case of SF_4 it should be a resonance hybrid of ($SF_3^+/F^- \leftrightarrow SF_4$ etc.) [8]. However, these methods can not illustrate correctly molecular structures bearing an unusual valency. Rundle earlier, then Musher, had proposed the notation of "hypervalent bonding" [9].

According to molecular orbital calculations, d-orbital participation is not important for hybridization; instead, only $3s$- and $3p$-orbitals are sufficient to make a decet and a dodecet in unusual molecules [10]. The chemical bonding used in the decet is called a three-center four-electron bonding, namely, the central atoms should have two types of chemical bonding, one is a three-center four-electron[3c-4e] bonding (or an axial bond), and the other comprises three sp^2 hybridized orbitals orthogonal to each other (or an equatorial bond); hence, the molecular structure should be a trigonal bipyramid (TBP). These hypothetical molecular structures are illustrated in Fig. 1.

In contrast, dodecet species have three sets of three-center four-electron bonds, hence, the corresponding molecule should have an octahedral structure (OC), as shown in Fig. 2 [8c, 10]. However, strictly speaking, these bonds differ

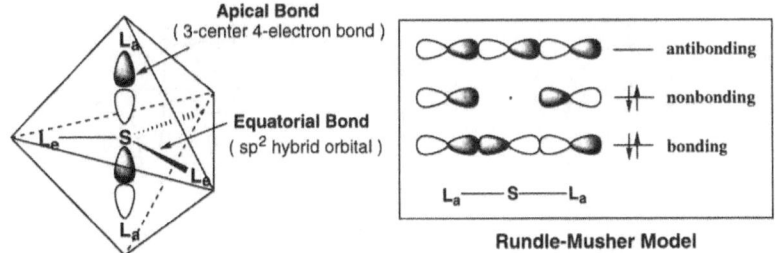

Figure 1. Trigonal bipyramidal structure and Rundle-Musher model of 3c–4e bond

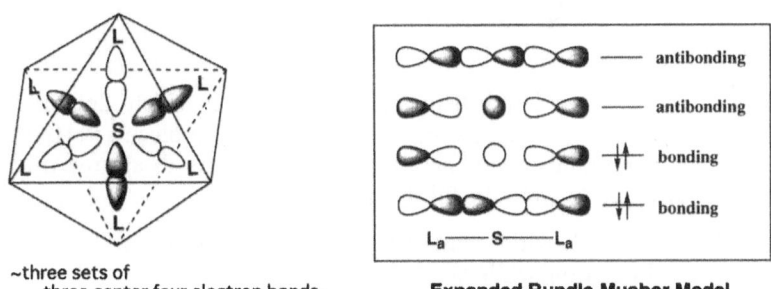

Figure 2. Octahedral structure and expanded Rundle-Musher model of 3c-4e bond

from those of decet species. When decet species change into dodecet species, the three sets of 3c-4e bonds of dodecet species contains a 3s orbital on the central atom, and the non-bonding orbital of the 3c-4e bond on the decet species is split into new bonding and anti-bonding orbitals (expanded Rundle-Musher model) [1k, 11]. Therefore, the corresponding 3c-4e bond is stronger than that of the decet species. Generally, the hexacoordinated chalcogen species are more stable than the corresponding tetracoordinated species [12].

In organic sulfur chemistry, these decet species have long been considered to be transition states or unstable intermediates in the nucleophilic substitution reaction (S_N2), and not to exist as real molecules except in a few cases [13]. However, recently numerous heteroatom compounds containing hypervalent structures have proved to be of considerable interest to many organic chemists.

The hypervalent molecules are now known for the elements of below the 3rd row and from the 14th to 17th groups in the periodic table. Among them, the chemistry of hypervalent compounds of the 16th group elements, the chalcogens, has been most widely investigated, particularly sulfuranes, while both the selenium and tellurium derivatives, called selenuranes and telluranes, have also attracted attention. Investigation of these compounds has been growing as a new field in organic chemistry. Since the pioneering work by Martin and Arhart [14] and Kapovits and Kalman [15] on the first preparation of sulfuranes 1 and 2 (Scheme 1), sulfuranes and persulfuranes have been extensively and systematically investigated [1a].

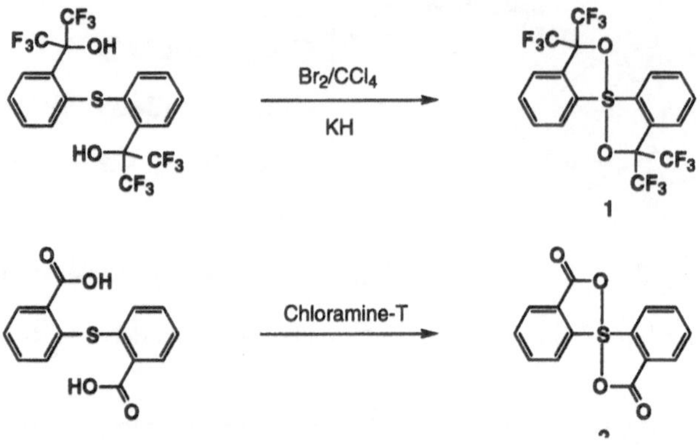

Scheme 1

Their chemical and physical properties have been widely documented by Martin and co-workers. According to Martin *et al.*, molecular structures of hypervalent compounds or the arrangement of the ligands around the central elements (TBP) conform to the following restrictions:

1. The Mutterties rule which states that the more electronegative ligands occupy the axial position, while electropositive ligands must be at the equatorial position; lone pair electrons should be located at the equatorial position.
2. Five- and six-membered rings which stabilize the hypervalent compounds occupy both the axial and equatorial positions [1a, 16].

Of the chemical properties of hypervalent compounds, there are two typical reactions; one is the ligand exchange reaction and the other is the ligand coupling reaction [1g, 17]. Pseudorotation or turnstile rotation [18] is also a typical stereochemical phenomenon observed in hypervalent compounds having pentacoordinated structures [19]. In the mechanism of nucleophilic substitution reactions at the sulfur, it has long been argued whether the reaction proceeds via a pentacoordinated species as an intermediate, unlike the S_N2 type substitution at carbon, or through a simple S_N2 type process. In these reactions, the nucleophiles must come from the apical direction and the leaving group should also be eliminated from the apical position, i.e., an a-a process resulting in inversion of stereochemistry. There are a few other possible combinations of attacking and leaving groups, namely a-e (or e-a) processes in the reactions, in which the stereochemical results are either retention or inversion, and hence quite different from that of an a-a process as shown in Fig. 3 [20].

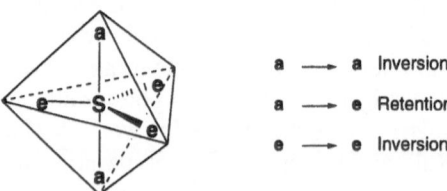

Figure 3. Stereochemical processes

In contrast to sulfur, selenium and tellurium derivatives seem to show slightly different behavior. In particular, tellurium gives various stable hypervalent compounds (telluranes), and hence it has become one of the most useful elements as a model for investigation of sulfuranes and selenuranes [1b,c].

According to molecular orbital calculations [8a], important factors to be noted for preparation of hypervalent compounds are as follows:

1. The smaller the electronegativity of the central element, the more stable the hypervalent compounds, that is, the three-center four-electron bonding must be stabilized
2. As to the ligand, the more electronegative elements or groups stabilize the hypervalent bond
3. Lone pairs of electrons must play an important role for formation and stabilization of the hypervalent sulfuranes

A combination of electropositive elements and electronegative ligands thus provides the most stable hypervalent bonding. In the case of chalcogenuranes, tellurium best fulfills these requirements and gives the most stable hypervalent molecules, followed by selenium and sulfur. As to the ligands, fluorine, chlorine, and oxygen must be appropriate candidates for the ligands together with their substitutes. As predicted from valence bond treatment by Shaik *et al.* [21], the lower the bond energy, the stronger the hypervalent bond formed. Telluranes are indeed known to be more stable than selenuranes, and sulfuranes are the least stable. The various essential properties of the chalcogen elements are summarized in Table 1 [22].

Table 1. Physical properties of chalcogen elements

| | Chalcogen atoms | | | |
	O	S	Se	Te
Electronic configuration	$2s^22p^4$	$3s^23p^4$	$3d^{10}4s^24p^4$	$4d^{10}5s^25p^4$
Electronegativity	3.5	2.5	2.4	2.1
Electroaffinity	7.28	–3.44	–4.21	–
Ionization potential	13.61	10.36	9.75	9.01
pKa (aq. MH_2)	16	7.0	3.8	2.6
pKa (aq. MH)	12.9	11.0	11.0	
Bond energy (kcal/mol)				
Ch–Ch	51.4	65.5	44,0	33.0
Ch–H	110.7	91.5	66.9	56.9
Ch–C	86.0	65.0	57.8	–
Bond length (Å)				
$C(sp^3)$–M	1.41	1.81	1.98	2.15
$C(sp^2)$=M	1.22	1.54	1.67	–

In organic sulfur chemistry, there is a great deal of indirect evidence to prove that sulfuranes are intermediates in substitution reactions. In addition, sulfuranes having electronegative ligands such as SF_4 [23], Ar_2SCl_2 [24], etc., are known, and, for example, *p,p'*-dichlorophenyl-dichlorosulfurane has been isolated and proved to have a TBP structure by X-ray crystallography.

In the reaction of triphenylsulfonium chloride with phenyllithium or Grignard reagents, the products obtained are biphenyl and diphenyl sulfide in quantitative yields. Earlier, the mechanism for this reaction was investigated by Franzen and Mertz [25], and they described the reaction as proceeding through initial formation of tetraphenylsulfurane, which had not previously been detected at that time. Similar treatment of the triphenylselenonium salt with phenyllithium by Wittig *et al.* also afforded biphenyl and diphenyl selenide, but the authors were unable to detect tetraphenylselenurane. However, tetraphenyltellurane was actually isolated as a crystalline compound which was shown to undergo various substitution reactions with nucleophiles in addition to ligand coupling reactions resulting in the quantitative formation of biphenyl and diphenyl telluride [26]. This is the first example of isolation of a tellurane having four carbon ligands. Subsequently, a more stable tellurane having four carbon ligands, bis(2,2'-biphenylylene)tellurane, was prepared by Hellwinkel *et al.*, and a new aspect of the chemistry of hypervalent compounds was thus advanced [27]. Hellwinkel *et al.* also isolated bis(2,2'-biphenylylene)selenurane [28] and, quite recently, Sato *et al.* have succeeded in synthesizing bis(2,2'-biphenylylene)sulfurane as stable crystals, and have determined the structure by X-ray crystallographic analysis [29]. They also succeeded in detecting the tetraphenylsulfurane and selenurane by a low temperature NMR technique [30]. These compounds are described in a subsequent section. It is now acknowledged that tetraarylchalcogenuranes are common intermediates for the substitution reactions of triarylchalcogenium salts with organometallic reagents. Since the chemistry of sulfurane has been well documented by Hayes and Martin [1a] and by Mikolajczyk *et al.* [1h], this review describes recent advances in the hypervalent compounds of sulfur. The article is divided into three parts: (1) sulfuranes as reactive intermediates; (2) recent advances in the study of stable sulfuranes together with some selenium and tellurium analogs as a comparison; (3) preparation of persulfuranes.

2
Sulfuranes as Reactive Intermediates

It has long been argued whether sulfuranes are actually formed as intermediates in the nucleophilic substitution reactions of organosulfur compounds or are merely transition states analogous to that for an S_N2 reaction at a carbon atom. Thus, two distinct processes are conceivable in the reactions; one is a simple one-step process and the other is constructed with at least two steps involving sulfurane as an intermediate. There are a number of reports treating this problem [1fg, 17, 31].

Dicoordinated sulfur species, such as disulfides, have been considered to be convenient molecules for the investigation of nucleophilic substitution reactions. However, there was no apparent evidence to distinguish the reaction pathway as a one- or two-step process.

Tricoordinated sulfur compounds, such as sulfoxides, sulfonium salts, and related derivatives, undergo nucleophilic substitution upon treatment with nucleophiles, and are useful for investigation of the mechanism at the reactions on the sulfur atom. Furthermore, optically active compounds are readily prepared in these derivatives which may provide additional evidence to aid understanding of the stereochemical processes in the substitution reactions. Although there remains little evidence that substitution at the tricoordinated sulfur atoms proceeds through initial formation of sulfurane, much indirect evidence has been accumulated that the reactions indeed proceed through formation of a sulfurane, from which either ligand exchange or a ligand coupling reaction takes place to afford the final products. Although many attempts have been undertaken to isolate or even merely detect the sulfurane in these reactions, only a few sulfuranes have been detected by spectroscopic methods [32].

A typical investigation to prove kinetically the formation of a sulfurane as an intermediate in the alkyl transfer reactions at the sulfonium sulfur has been demonstrated by Young and co-workers [33]. They used S-substituted benzyl-S-methyl-S-(substituted)phenylsulfonium salts (3) with amines as nucleophiles. The Hammett $\varrho\sigma$ relationship and the β-values of the amines suggest that the reactions are involved in equilibrium formation of a sulfurane 4 from which ligand coupling takes place to give the final N-benzylamines and thioanisole derivatives, as shown in Scheme 2.

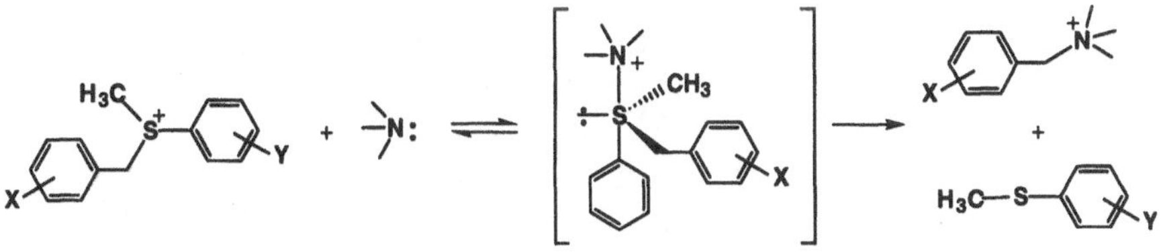

3   **4**

Scheme 2

As indirect evidence for the formation of sulfuranes, the following reactions have been reported by Okuyama and co-workers [34]. They undertook the hydrolysis of sulfinic esters and sulfine amides 5 under acidic conditions using an ^{18}O-tracer. On the basis of ^{18}O exchange and kinetic studies for the hydrolysis, they found that ^{18}O exchange takes place in the reactions, and proposed that the mechanism for the hydrolysis involves the initial formation of sulfurane 6 as an intermediate as shown in Scheme 3.

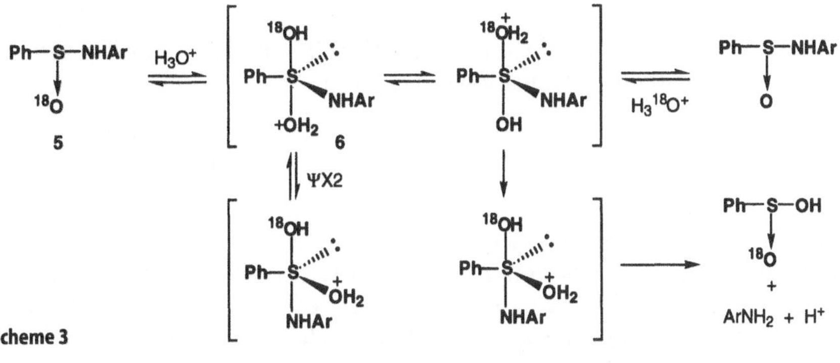

Scheme 3

2.1
Reactions of Sulfoxides with Organometallic Reagents

Sulfoxides react with organolithium and Grignard reagents by four different reaction pathways. One involves nucleophilic substitution with Walden inversion at the sulfinyl sulfur atom. A second involves a concomitant ligand exchange and disproportionation reactions. A third process, recently discovered, comprises ligand coupling reactions, and the fourth is the formation of α-sulfinyl carbanions by proton abstraction. The four reactions are summarized in Scheme 4.

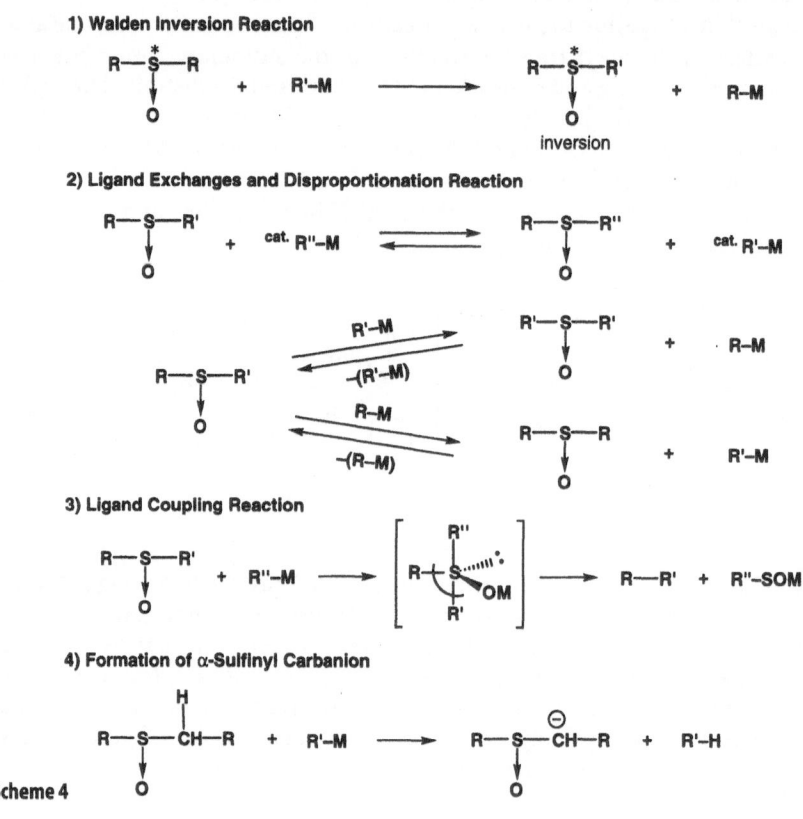

Scheme 4

These reactions are presumed to proceed via an initial formation of sulfuranes as intermediates or as transition states [1f, g].

2.1.1
Ligand Exchange Reactions

Sulfoxides undergo facile ligand exchange reactions upon treatment with organolithium or Grignard reagents in ether. In the reactions using alkyl aryl or haloalkyl aryl sulfoxides, the most electronegative ligand is usually replaced by the organometallic reagent to give dialkyl sulfoxides with complete inversion at the sulfur atom. This reaction is thus used for the synthesis of optically active dialkyl sulfoxides which are difficult to prepare by conventional processes [35] (Scheme 5).

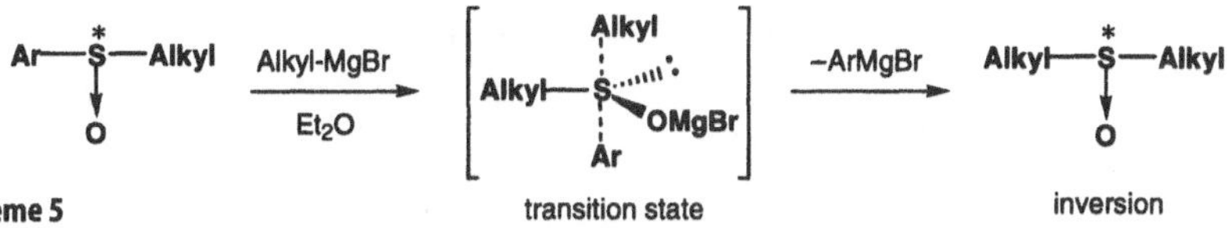

Scheme 5   transition state   inversion

On the basis of the stereochemical results, these substitution reactions are considered to proceed by a simple S_N2-type process involving a hypervalent sulfurane as a simple transition state, and not an intermediate.

When optically active methyl p-tolyl sulfoxide is treated with MeLi, the racemization of the sulfoxide has been observed, indicating facile elimination of p-tolyl group and rapid recombination to the sulfoxide, proposed by Mislow et al. [36], as shown in Scheme 6.

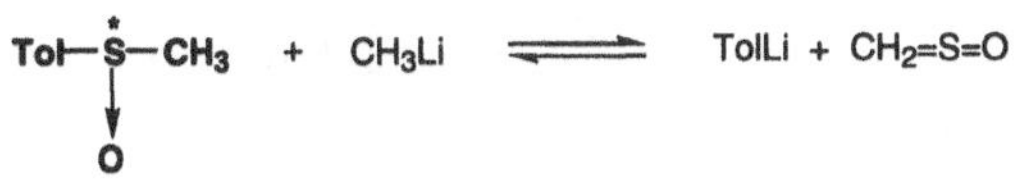

Scheme 6

Similar reactions of sulfoxides bearing a t-butyl group as a ligand were also found to give racemization products [35b]. The mechanism in this case was considered to proceed through an electron transfer process.

Previously, Andersen et al. had attempted to prepare sulfuranes from the reactions of diaryl sulfoxides with aryl Grignard reagents and aryl lithium agents. They reported that the reactions give initially the corresponding triaryl sulfonium salts, from which two pathways, namely the formation of tetraaryl sulfuranes and the formation of benzyne, take place simultaneously, on the basis that they obtained diaryl sulfides and biaryls derived from benzyne, as shown in Scheme 7 [37].

However, Trost et al. treated triaryl sulfonium salts with aryl lithiums to obtain only diaryl sulfides and diaryls, via sulfuranes [38]. These two contradictory results obtained by different groups have not been resolved.

As an example of the few exceptions to the normal pathways of the reactions of sulfoxides with organometallic reagents, earlier Gilman and Swayampatai found that thianthrene monosulfoxide **9** reacts with BuLi to afford the ring contracted dibenzothiophene **10** in low yield as shown in Scheme 8 [39]. This reaction has been applied to synthesize various 1,9-disubstituted dibenzothiophenes **11** and selenophenes from the reactions of 4,6-dithiosubstituted thianthrenes and selenanthrenes with organometallic reagents [40]. The reaction mechanism has been postulated to proceed through initial formation of sulfurane **12**, or the corresponding selenurane, as an intermediate which is greatly stabilized by the two arylthio groups at the 1,8-positions of thianthrene (Scheme 8).

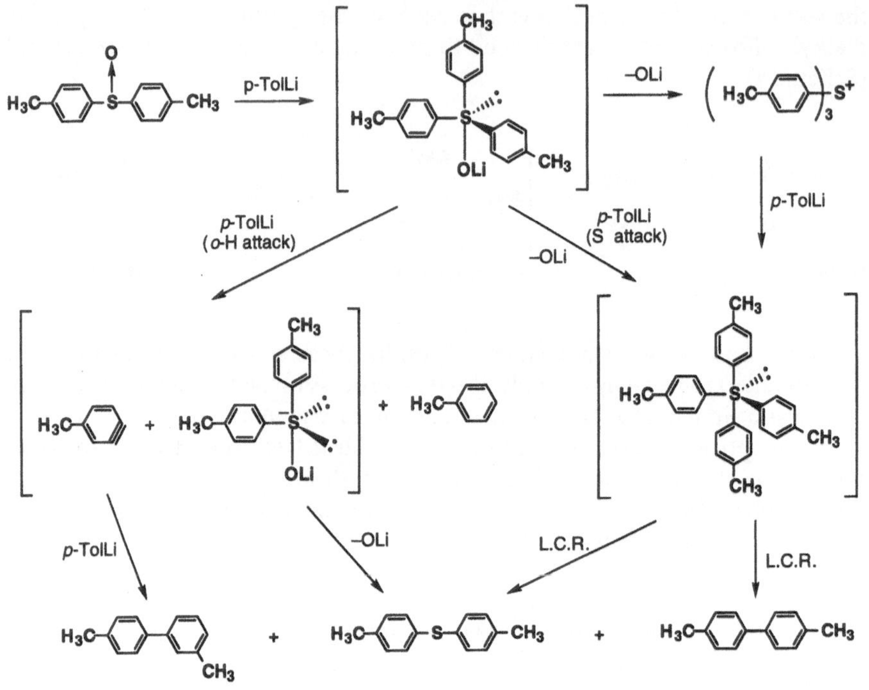

Scheme 7   L. C. R. = ligand coupling reaction

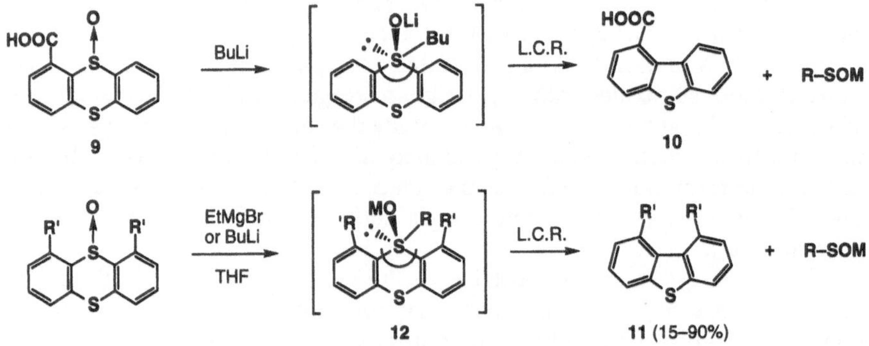

(R = Et, Bu; R' = MeS, PhS; M = Li, MgBr)

Scheme 8   L. C. R. = ligand coupling reaction

Similarly, Potratz *et al.* found that dibenzothiophene monoxide reacts with aryl Grignard reagents to give aryl dibenzothiophenium salts **13** in moderate yields. The mechanism for the reaction involves a simple S_N2 type reaction at the sulfur atom [41], in which the oxygen atom becomes a leaving group (Scheme 9). There are only a few reactions where the oxygen atom is a leaving group while alkoxy sulfonium salts react with many organometallic salts as nucleophiles to afford sulfonium salts in good yields. The reactions of sulfoxides or seleno-

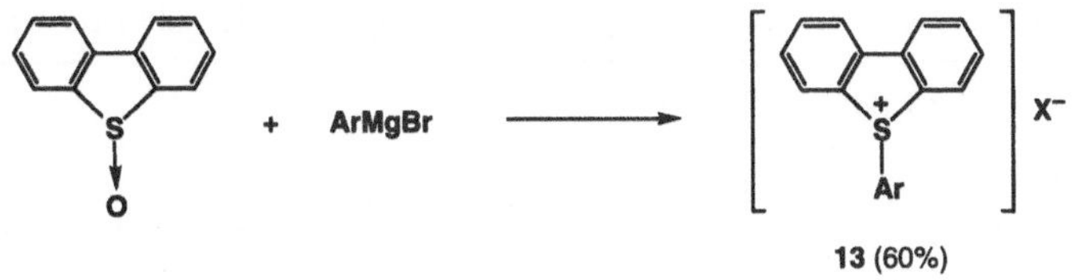

Scheme 9 13 (60%)

xides in the presence of trimethylsilyltriflate (Me$_3$SiOSO$_2$CF$_3$) give sulfonium and selenonium salts in one pot [42] (Scheme 10).

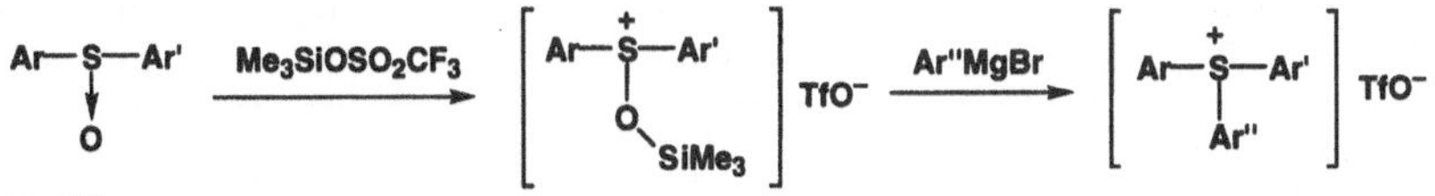

Scheme 10

Interestingly, when phenyl p-tolyl sulfoxide is treated with n-BuLi, facile ligand exchange reaction takes place to give the three sulfoxides by disproportionation in a statistical ratio of 1:2:1 even at –110 °C. This disproportionation reaction is observed generally in diaryl sulfoxides upon treatment with small amount of alkyl or aryl lithium reagents, e.g., less than 0.1 molar equivalent of BuLi. This reaction takes place rapidly at from –80 °C to around –110 °C to give the three different sulfoxides, soon after mixing the sulfoxides with BuLi, with complete racemization [43]. The results are shown in Scheme 11 and Table 2.

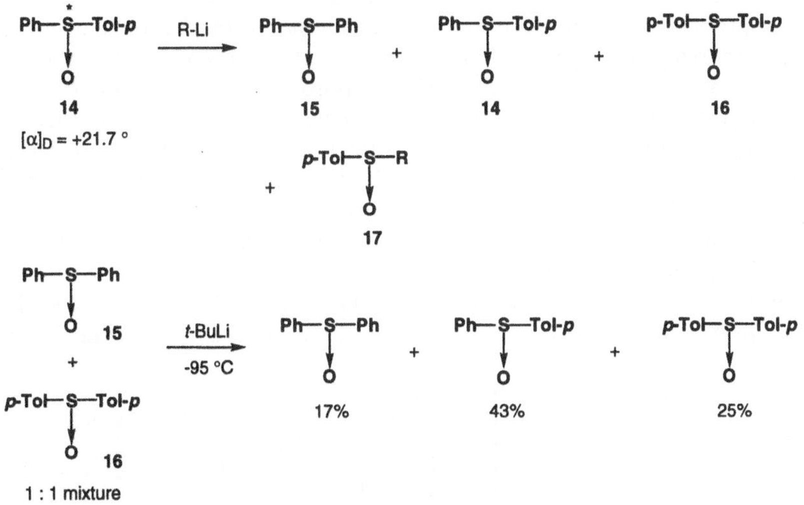

Scheme 11

An equivalent mixture of diphenyl **15** and di-p-tolyl **16** sulfoxides also gave similar disproportionated sulfoxides in a 1:2:1 ratio.

This rapid ligand exchange reaction was observed not only in simple aryl sulfoxides but also bulky sulfoxides with racemization of the recovered sulfoxide. The reaction is initiated by attack of organolithium reagent on the sulfinyl sul-

Table 2. Products from the reaction of optically pure phenyl p-tolyl sulfoxide with RLi

R	Molar equiv.	Yield (%)				$[\alpha]_D$	
		14	15	16	17	15	17
n-Bu	2.0	0	0	0	20[a]	–	0
t-Bu	2.0	0	0	0	30[b]	–	0
t-Bu	1.0	10	29	21	24	0	0
t-Bu	0.5	14	35	22	15	0	
t-Bu	0.2	17	41	–21	7	0	
Ph	0.2	24	48	22	0	0	
LDA	1.0	0	100	0	0	20.3	

[a] $(n\text{-Bu})_2SO$, 55%.
[b] $t\text{-BuS(O)SBu-}t$, 21%.

fur atom, with elimination of one aryl group as a ligand exchange product. This ArLi, once formed, attacks the starting sulfoxide to form a sulfurane, from which another ligand is extruded. These exchange processes are repeated, even at from –85 °C to around –110 °C, to give a mixture of disproportionation products. Furthermore, interestingly, the initially substituted n-butyl p-tolyl sulfoxide is also completely racemized. These results suggest that the reaction is not a simple S_N2-like substitution as described previously. These facile ligand exchange reactions have never been previously observed, and contrast markedly with the results described by Andersen *et al.* [37]. The difference between these two reactions is probably due to the difference in the temperature and concentrations of the lithium reagents employed in the reactions. At lower temperature, the initial step for the substitution takes place by an approach of RLi group with coordination to the sulfinyl oxygen atom from the back side of the more electronegative aryl group, to form an unstable sulfurane that undergoes rapid pseudorotation and then gives ligand exchange products, namely, racemic n-butyl p-tolyl sulfoxide and PhLi. At the higher temperatures employed by Andersen, however, RLi may approach from the back side of the OLi, group to displace OLi which should be a better leaving group than the aryl group. In sulfinyl compounds, the C-S bond energy (60–70 kcal) is less than that of the S-O bond (85–90 kcal), so that at lower temperatures the C-S bond is cleaved preferentially, while at higher

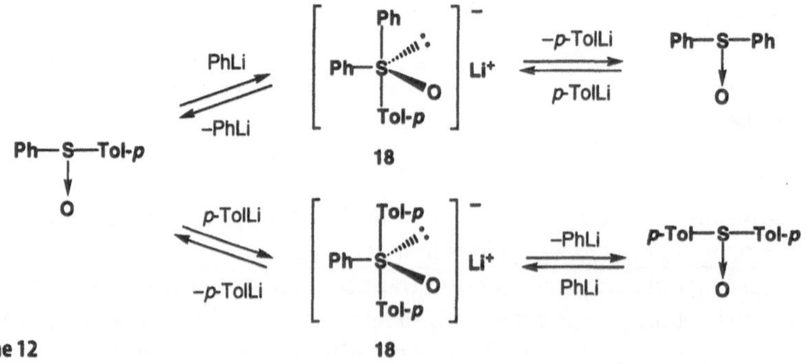

Scheme 12   18

temperatures S-O bond cleavage should also take place more readily than C-S bond cleavage due to the preferential attack by the alkyl lithium from the back side of the S-O group [44]. The facile aryl group exchange process can therefore be rationally explained in terms of the formation of sulfuranes 18, as shown in Scheme 12.

As an extension to this facile ligand exchange reaction, dibenzothiophene monoxide reacts with catalytic amounts of alkyl lithium reagent to afford oligomers of o-biphenylylene sulfide, via a dimer 19, as shown in Scheme 13 [45].

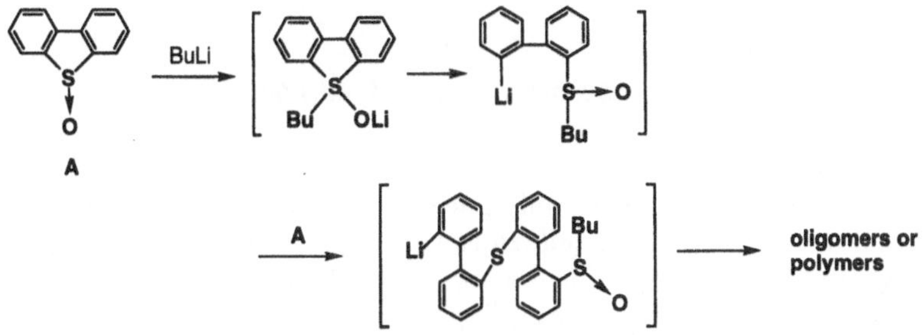

Scheme 13 19

Similarly, diaryl selenoxides also undergo facile ligand exchange and disproportionation reactions. However, selenoxides react more readily than sulfoxides. For example, phenyl p-tolyl selenoxide reacts with 0.5 equivalents of t-BuLi (or PhLi) to give both a mixture of three selenoxides and three reduced selenides in ca. 1:2:1 ratio, with concomitant formation of diaryls, as shown in Scheme 14 and Table 3 [46].

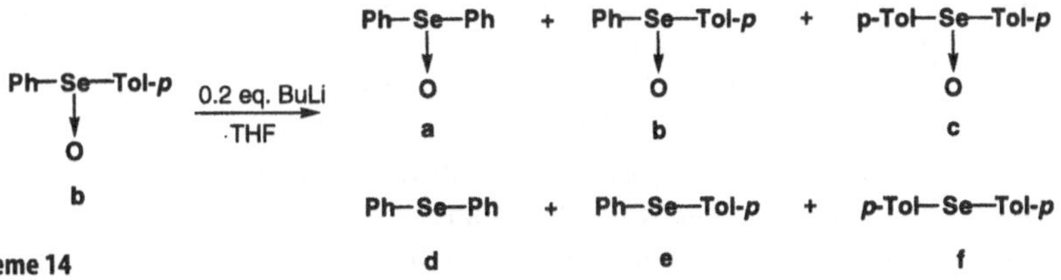

Scheme 14

Table 3. Products from the reaction of phenyl p-tolyl selenoxide with RLi

RLi	Equiv.	Temp (°C)	Yield (%)					
			a	b	c	d	e	f
n-BuLi	0.2	−78	17	31	22	2	2	1
n-BuLi	1.0	−78	4	13	9	5	10	5

The ratio of these ligand exchange and reduction products also depends largely upon the temperature and molar ratio of the RLi reagents. In the selenoxides, the Se-O bond energy is lower than that of S-O, and hence Se-O bond cleavage should take place concomitantly with C-Se bond cleavage.

In order to confirm the formation of these hypervalent compounds directly, ^{1}H and ^{13}C NMR spectra of the reaction mixtures of sulfoxide or selenoxide with PhLi were measured at -80 °C. However, neither sulfurane nor selenurane was detected. The spectra were consistent with those of a mixture of disproportionated sulfoxides (selenoxides) and aryl lithiums [46].

2.1.2
Ligand Coupling Reactions of Sulfoxides Bearing Pyridyl Groups

Although there are many examples of ligand exchange reactions of sulfoxides with organometallic reagents, only a few ligand coupling reactions have been reported in the reactions of sulfoxides with organometallic reagents, until the fairly recent investigations of heteroaryl sulfoxides reported mainly by Oae and his co-workers [1e, 47]. Ligand coupling is a well-known reaction of hypervalent species [1j]. In general, ligand coupling is considered to take place between two electronegative ligands, according to the theoretical MO calculation in which two apical ligands undergo coupling in a concerted manner [1a, 48]. However, this reaction mode requires pseudorotation or turnstile rotation to allow the ligands to come sufficiently close together to satisfy stereoelectronic requirements. In this process, the stereochemistry of the carbon atoms attached at the coupling center involves retention of configuration, indicating that the reaction proceeds by a concerted front-side attack of one ligand on the other via a three-membered transition state or intermediate, without bond breaking. Thus, if there is any cohesive interaction between the two ligands, they would be extruded from the central valence-shell expanded sulfur atom in a concerted manner, affording a ligand-coupling product in which the original configurations of the two coupled ligands are retained completely. In the reactions of sulfoxides with organometallic reagents, ligand exchange is a general phenomenon, while ligand coupling has been observed only in a few structurally restricted sulfoxides. These results suggest that the reactions of sulfoxides undergo ligand exchange via sulfuranes as transition states and not intermediates. As shown in the following section, many sulfonium salts undergo ligand coupling reactions, and not ligand exchange reactions, upon treatment with organometallic reagents, and, in the case of sulfonium salts bearing strongly electronegative substituents, the intermediate sulfurane has actually been detected. This result suggests that sulfoxides bearing electronegative ligands can undergo ligand coupling reactions through the formation of sulfuranes as intermediates. Thus one can postulate that the ligand coupling reaction is a diagnostic procedure for the intermediacy of sulfuranes. The first example of ligand coupling reactions, now a generally accepted phenomenon, is the reaction of benzyl pyridyl sulfoxide with organometallic reagents giving 2-benzylpyridine together with sulfenate derivatives, which, after treatment with CH_3I or hydrolysis, affords sulfoxide or thiolsulfinate in high yields [49].

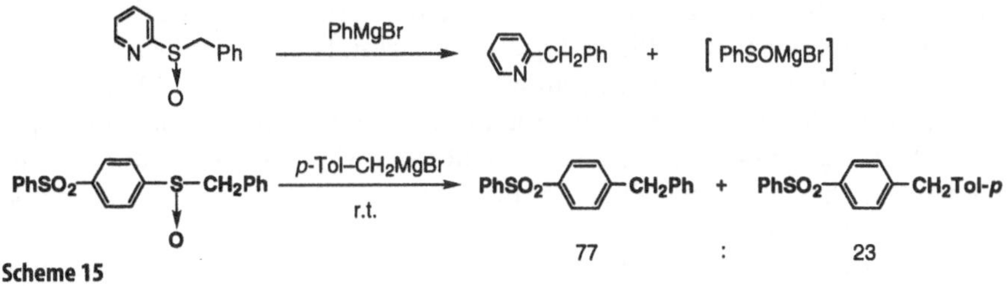

Scheme 15

Obviously, this reaction is not the same as the simple ligand exchange reactions described in the previous section. The reaction seems to proceed along a similar pathway to that of sulfonium salts with organometallic reagents described in the following section. The reaction, as shown in Scheme 15, has been found in many other sulfoxides bearing at least one electron withdrawing heteroaryl group, such as 2-pyridyl-, 2-quinolyl and their 4-isomers. As the second ligand, not only benzyl but also many other substituents, such as allylic [50], vinylic-, [51] sec- and tert-alkyl groups, and aryl groups, can also couple with the heteroaryl groups to produce substituted heteroarylic compounds in high yields [1g]. Even a phenyl group bearing an electron-withdrawing group, such as p-phenylsulfonylphenyl, can also replace the heteroaryl group and achieve a smooth coupling reaction [52].

Ligand coupling therefore becomes a general phenomenon in the reactions of sulfoxides bearing heteroaryls with Grignard or organolithium reagents [1j]. Grignard reagents are better to use than organolithium reagents, because organolithium reagents work as both thiophilic reagents and bases toward heterocyclic compounds, and can give a complex mixture of intractable products. As a typical example, benzyl 2-pyridyl sufoxide reacts with PhMgBr in THF at room temperature to give 2-benzylpyridine in nearly quantitative yield, whereas

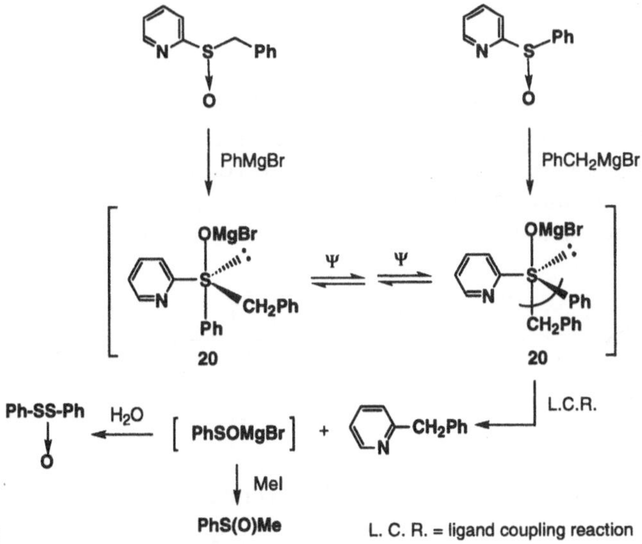

Scheme 16   L. C. R. = ligand coupling reaction

phenyl 2-pyridyl sulfoxide also affords 2-benzylpyridine in a nearly identical yield upon treatment with PhCH₂MgCl. These results clearly indicate that the coupling reactions proceed via a common intermediate, sulfurane **20**, which can undergo pseudorotation or turnstile rotation. The remaining organic sulfur species is PhSOMgX, which can be converted into methyl phenyl sulfoxide or can be quenched with water to give PhS(O)SPh(via PhSOH) and its disproportionation products, as shown in Scheme 16.

A cross-over experiment using an equimolar combination of two different sulfoxides was carried out. Thus, a mixture of benzyl 2-pyridyl sulfoxide **21** and (4-methylphenyl)methyl 6-picolyl sulfoxide **22** was treated with PhMgBr, giving exclusively 2-benzylpyridine and 2-(4-methylphenyl)-6-methylpyridine in good yields. None of the cross-over products was obtained at all, establishing that the coupling reaction must proceed intramolecularly [49]. The formation of sulfurane **20** is thus a rational explanation for the reaction mechanism. Because the coupling was found to proceed nearly quantitatively, a stereochemical study of the coupling reaction was carried out using optically active 1-phenylethyl 2-pyridyl sulfoxide **23**.

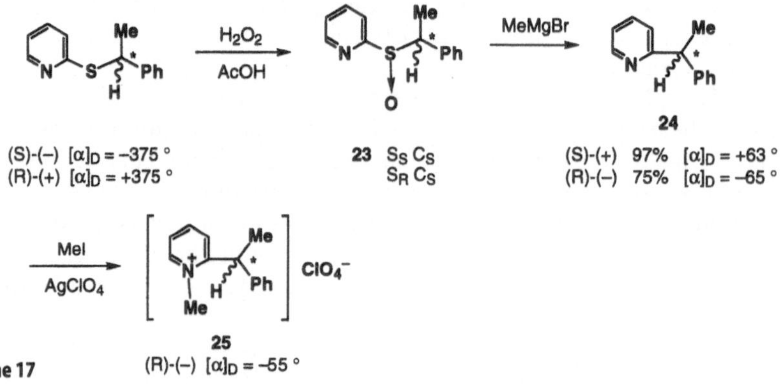

Scheme 17

The results, summarized in Scheme 17, clearly demonstrate that the configuration at the migrating carbon center in the sulfoxide is completely retained after the coupling reaction takes place. The configuration of 2-(1-phenylethyl)pyridine **24** was confirmed by X-ray crystallographic analysis after conversion into the crystalline *N*-methylpyridinium salt **25** [49].

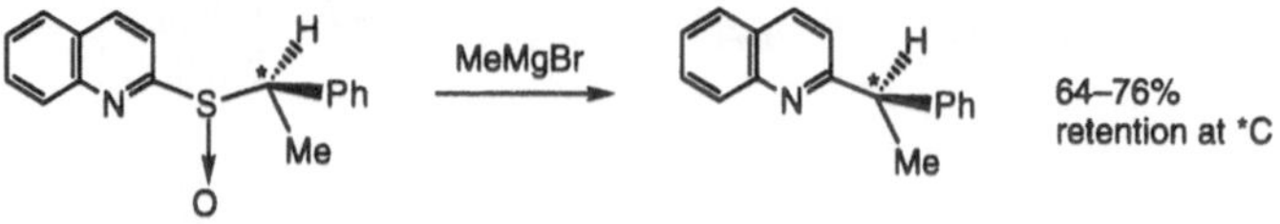

Scheme 18

Furthermore, Oae and his co-workers have conducted similar stereochemical reactions using optically active (1*R*)-phenethyl 2-quinolyl(*R*-) or (*S*-)sulfoxide

26 with MeMgBr, and found that the configuration of the coupling product (1R)-2-(1-phenethyl)quinoline 27 obtained was retained (Scheme 18).

Recently, retention of configuration during the coupling reactions of sulfoxides has also been reported by Neef *et al.* [53] and Okamura and Teobald [54].

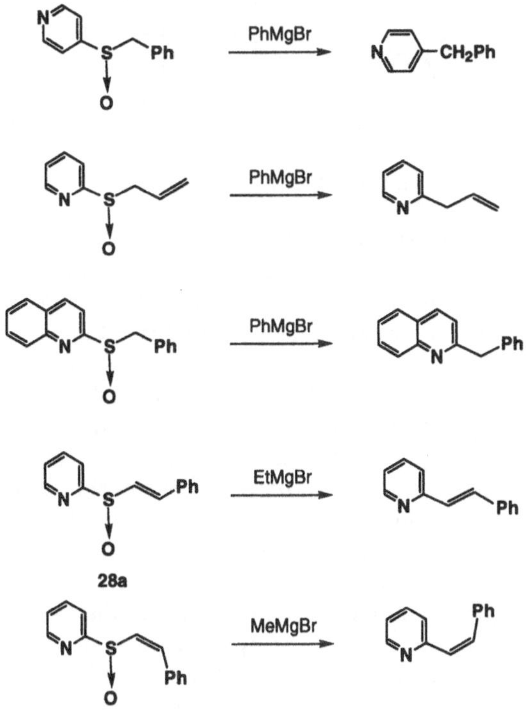

Scheme 19   28b

Coupling reactions with geometrical isomers of 2-pyridyl styryl sulfoxides 28a,b were also carried out, and each coupling product obtained was found to be produced with complete rentention of stereochemistry at the vinylic positions, as shown in Scheme 19.

All these stereochemical results strongly support the contention that the ligand coupling reactions of sulfoxides bearing electron withdrawing substituents proceed through the mechanism shown in Scheme 16, involving sulfuranes as intermediates.

Besides these intramolecular ligand coupling reactions of benzyl 2-pyridyl sulfoxides, it was found that methyl and other alkyl 2-pyridyl sulfoxides 29 undergo initial ligand exchange reaction to give 2-pyridyl Grignard reagent 30, which further attacks the starting sulfoxide to give 2,2'-bipyridyl through a ligand coupling reaction, in high yields. The Grignard exchange reaction was confirmed by a trapping experiment in the presence of benzaldehyde to give 1,1-(2-pyridyl)phenylethyl alcohol [55]. This cross-coupling reaction can be used for preparation of symmetrical 2,2'-bipyridyl derivatives, as shown in Scheme 20.

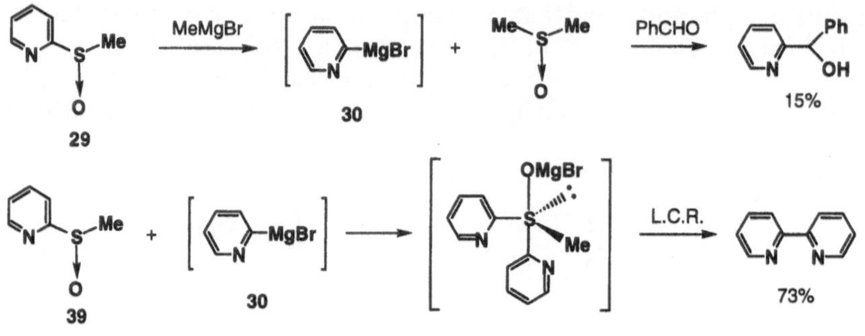

Scheme 20   L. C. R. = ligand coupling reaction

Phenyl 2-pyridyl sulfoxide gives different types of products upon changing the temperature and the Grignard reagents. At lower temperature (e.g., –78 °C) with ArMgBr, this sulfoxide gives only the ligand-exchange product. With *p*-TolMgBr, for example, at low temperature a mixture of phenyl and *p*-tolyl pyridyl sulfoxides was obtained. At room temperature, *p*-TolMgBr gives ligand-coupling products, namely, a mixture of 2-phenyl and 2-*p*-tolylpyridine and 2,2′-bipyridyl [56]. This observation result suggests that both the coupling and cross-coupling reactions take place simultaneously. When one uses EtMgBr at room temperature, the products are again both 2-phenylpyridine and 2,2′-bipyridyl. These different reaction modes of phenyl 2-pyridyl sulfoxide with Grignard reagents can be explained in terms of the formation of a Mg metal complex by the 2-pyridyl nitrogen atom taking the two equatorial positions, and thus the *p*-tolyl group can approach only from the back side of the electronegative phenyl group, to form a sulfurane. At higher temperature, the coordination of MgBr by the ligands is diminished, and hence the pyridyl ring must be placed at the equatorial position when the *p*-tolyl group approaches the central sulfur atom.

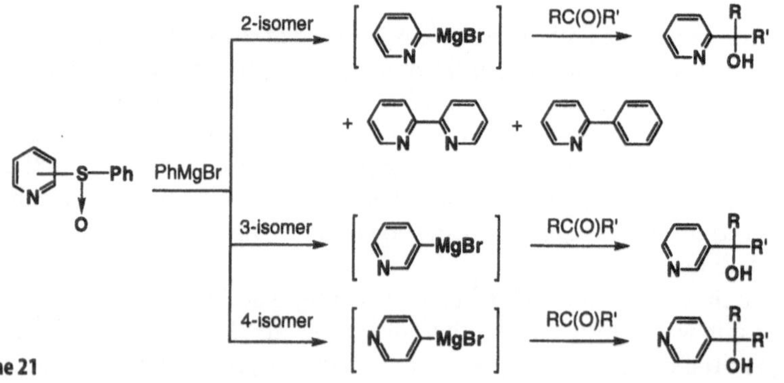

Scheme 21

In contrast to 2-pyridyl sulfoxide, both 3- and 4-pyridyl *p*-tolyl (or phenyl) sulfoxides were found exclusively to undergo the ligand-exchange reactions to eliminate stereospecifically 3- and 4-pyridyl Grignard reagents and phenyl *p*-tolyl sulfoxide. The optical activity of phenyl *p*-tolyl sulfoxide indicated that the stereochemistry had been completely inverted. The resulting 3- and 4-PyMgBr

Table 4. Ligand coupling and exchange reactions of phenyl pyridyl sulfoxides with PhMgBr

Sulfoxides[a]	Addn	Products[a]	Yield (%)
2-Py	PhCHO	2-Py–Py	31
		2-Py–Ph	17
		2-Py–CH(OH)Ph	13
3-Py	PhCHO	3-PyCH(OH)Ph	88
	α-NaphCHO	3-PyCH(OH)Naph-α	80
4-Py	PhCHO	4-PyCH(OH)Ph	64
	α-NaphCHO	4-PyCH(OH)Naph-α	63
	PhCH=CHCHO	4-PyCH(OH)CH=CHPh	60

[a] Py = pyridyl.

can be trapped with several electrophiles, such as aldehydes and ketones [57]. Some examples are shown in Scheme 21 and Table 4.

These results indicate that chelation of Mg by both the pyridyl nitrogen and the sulfinyl oxygen atoms play an important role in determining the stability of the intermediate sulfurane involving pyridyl rings.

As an extension to the ligand-coupling reactions, cross-coupling reactions using various phenyl pyridyl sulfoxides with pyridyl Grignard reagents were carried out, and the results are shown in Scheme 22 and Table 5 [58].

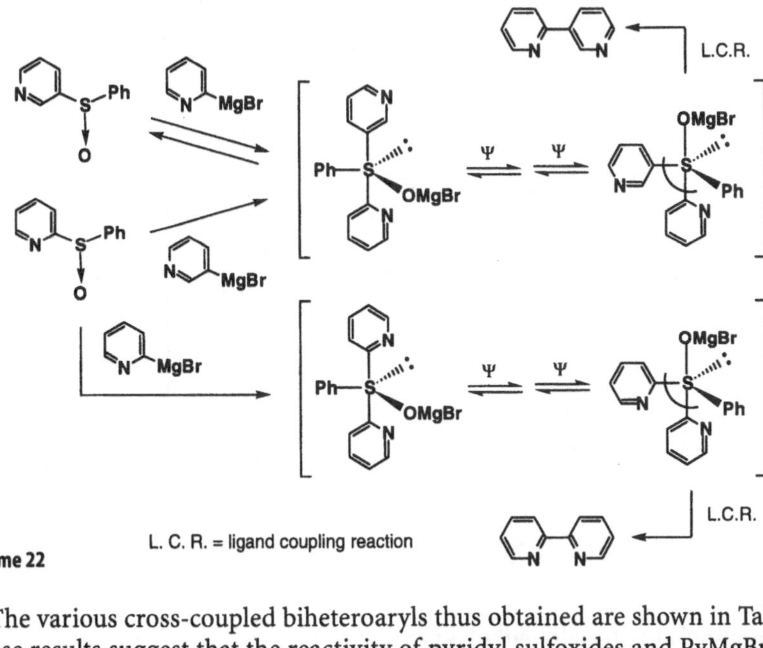

L. C. R. = ligand coupling reaction

Scheme 22

The various cross-coupled biheteroaryls thus obtained are shown in Table 4. These results suggest that the reactivity of pyridyl sulfoxides and PyMgBr is in the following order, namely, 2-Py > 4-Py ≫ 3-Py. This cross-coupling reaction is a convenient procedure for providing numerous unsymmetrical biheteroarylic compounds.

Table 5. Cross-coupling reactions of pyridyl grignard reactions and pyridyl sulfoxides

PyMgBr	Sulfoxide	Products	Yield (%)
2-Py	2-PyS(O)Ph	2-Py-Py-2	75
	3-PyS(O)Ph	2-Py-Py-2	23
		2-Py-Py-3	37
	4-PyS(O)Ph	2-Py-Py-4	58
3-Py	2-PyS(O)Ph	2-Py-Py-3	63
	3-PyS(O)Ph	no reaction	–
	4-PyS(O)Ph	2-Py-Py-4	34
		4-Py-Py-4	14
4-Py	2-PyS(O)Ph	2-Py-Py-4	63
	3-PyS(O)Ph	3-Py-Py-4	13
		4-Py-Py-4	25
	4-PyS(O)Ph	4-Py-Py-4	50

Furthermore, these ligand coupling and exchange reactions can be used for the preparation of unsymmetrically substituted bipyridyls and phenylpyridine derivatives. In order to introduce substituents in the pyridine ring, phenyl pyridyl sulfoxides were treated with lithium diisopropylamide (LDA) to give regiospecific lithiation in the pyridine ring of the sulfoxides. Upon treatment with electrophiles, the lithio derivatives give various o-substituted pyridyl phenyl sulfoxides. From these compounds the sulfinyl group can be removed, or cross ligand coupling reactions effected, to afford numerous pyridine derivatives bearing functional groups at appropriate positions or bipyridyl derivatives [59]. Several examples are collected in Scheme 23.

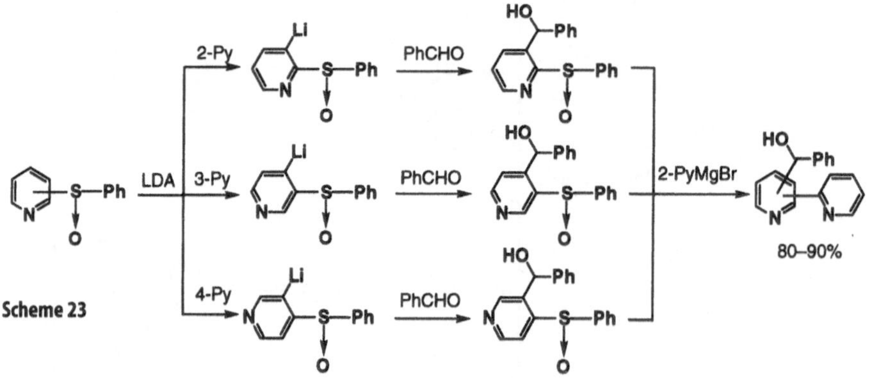

Scheme 23

3
Stable Organosulfurane Species

3.1
Symmetrical Spiro-Oxysulfuranes [10–S–4]

It is well known that stable spiro-sulfuranes require one or two sets of five-membered rings involving strong electronegative atoms at each apical position. Since most of the corresponding spiro-sulfuranes have already been discussed in many review articles, [1] this section focuses attention on the recent research with regard to the new symmetrical spiro-sulfuranes.

Livant *et al.* have reported the sulfurane having two axial oximate ($R_2C=NO$) ligands, as shown in Scheme 24 [60]. This sulfurane **31** was synthesized by the reaction of 2,2'-thiobis(acetopohenone) bis-oxime with *tert*-butyl hypochlorite

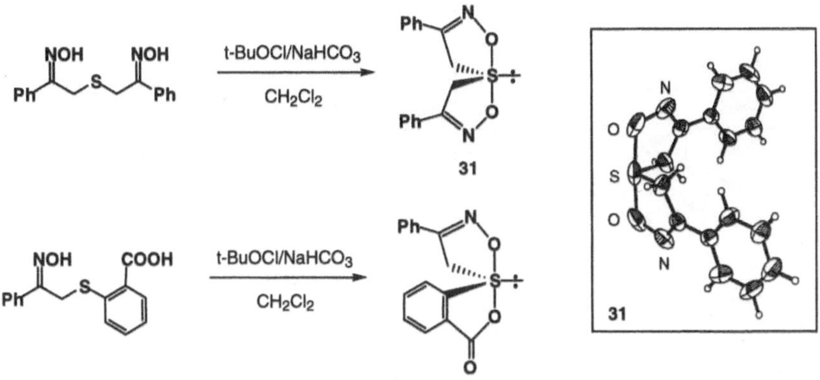

Scheme 24

in the presence of dry NaHCO$_3$ in dry CH$_2$Cl$_2$, and its structure was determined by X-ray crystallographic analysis, revealing that it has a trigonal bipyramidal geometry bearing two oximate groups as axial ligands. The authors have also synthesized the similar but unsymmetrical oximate sulfurane **32** under the same conditions.

Furukawa *et al.* have reported one of the most convenient methods to prepare various Martin's type sulfuranes as shown in Scheme 25 [61]. Thianthrene *S*-monoxide (**33**) was found to react with alkyl Grignard reagents affording the *o,o'*-bis-Grignard reagent of diphenyl sulfide, which was converted initially into the corresponding diols **34** on treatment with aldehydes or ketones, and then to the Martin's sulfuranes **35** by treatment with *tert*-butyl hypochlorite.

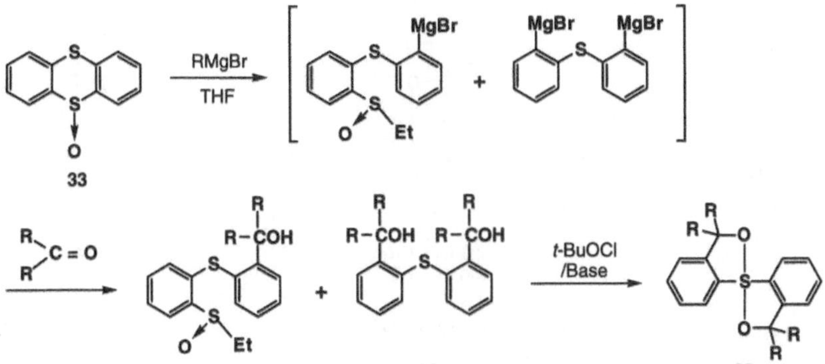

Scheme 25

Akiba *et al.* have provided tetracoordinated sulfuranes by the demethylation of a series of sulfur compounds such as **36**, having an equatorial methyl ligand,

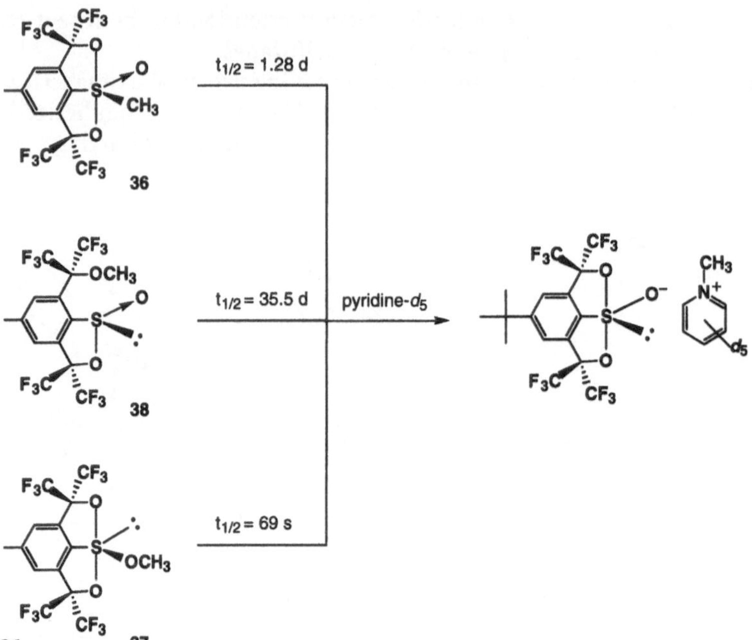

Scheme 26

37, an equatorial methoxy ligand, and sultine **38**, having a methoxy group, as shown in Scheme 26 [62]. These sulfuranes are all demethylated in pyridine-d_5 with a wide range of reaction rates.

Furthermore, two different reaction modes were found in sulfuranes **39** and **40**, having an equatorial methyl ligand, on treatment with an aqueous alkali solution in pyridine. The sulfurane **39** (R=CH$_3$) underwent a complete H-D exchange of the S-methyl group finally to give the S-CD$_3$-substituted sulfurane **41**. In contrast, the sulfurane **40** (R=CF$_3$) afforded the unexpected demethylation product, N-methylpyridinium sulfuranide **42**, under the same conditions, as shown in Scheme 27.

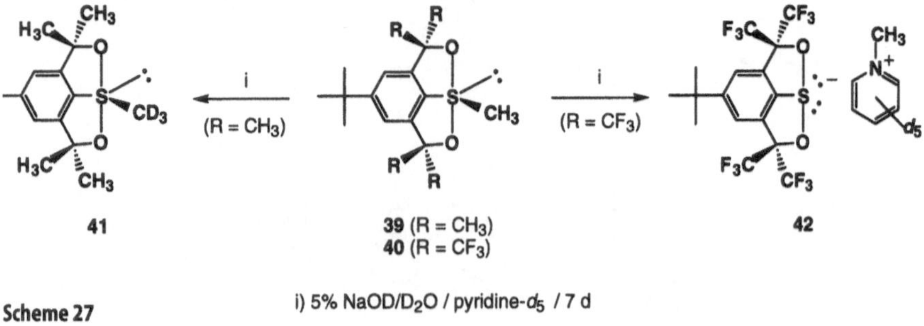

Scheme 27   i) 5% NaOD/D$_2$O / pyridine-d_5 / 7 d

Rongione and Martin have reported that the kinetic acidity of the protons in the equatorially substituted S-methyl group of the sulfurane oxide **43** is much reduced when the four methyl groups adjacent to the apical oxygens are repla-

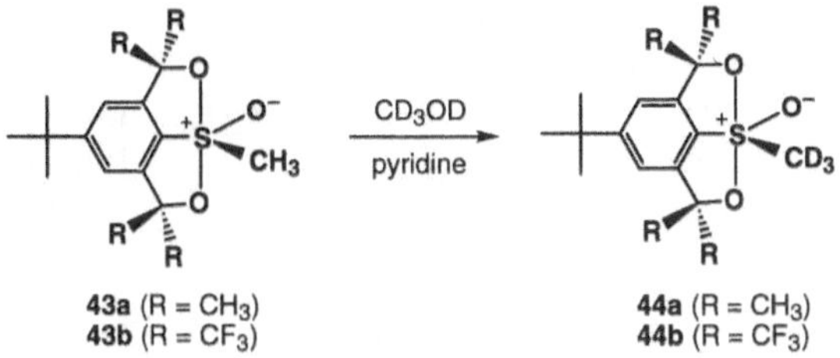

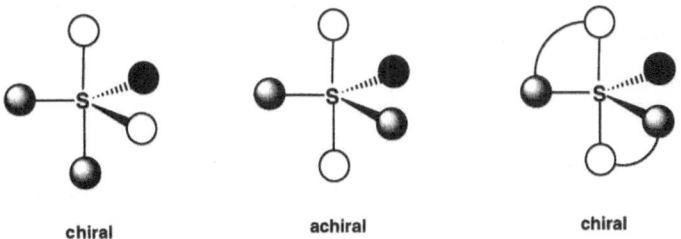

Scheme 28

43a (R = CH₃)
43b (R = CF₃)

44a (R = CH₃)
44b (R = CF₃)

ced by trifluoromethyl groups, as shown in Scheme 28 [63]. Deuterium exchange rates are faster for the CH_3-substituted compound **44a** than for the CF_3-substituted compound **44b** by a factor of ca. 86,000 at 25 °C.

Many spiro-oxysulfuranes have been already reported by Martin and his coworkers. Recently, tetra- and pentacoordinated spiro-sulfuranes have been reported, and their characteristic features with regard to their optical properties, dynamic behavior, and reactivities have attracted attention. It should be noted that these compounds become chiral when they contain at least three different ligands. However, due to the topological properties of such molecules, the spirosystems containing two pairs of equivalent substituents (unlike the acyclic analogs) are chiral, as shown in Fig. 4.

chiral   achiral   chiral

Figure 4. The topology of trigonal bipyramidal configurations of [10–S–4] or [10–S–5] type sulfuranes (● = lone electron pair or oxygen atom)

Allenmark and Claeson have reported that the enantiomers of a symmetric spirosulfurane **2** have been obtained by direct resolution of the racemate by chiral liquid chromatography, and further characterized by chiroptical methods [64].

Efficient synthetic approaches to the optically active spiro-sulfuranes and their oxides have been reported by Martin and Drabowicz [65]. The preparation of optically active spirosulfuranes **45** and **46** was performed by asymmetric dehydration of the corresponding prochiral sulfoxide diols **47**, as shown in Scheme 29. The optically active oxides **48** and **49** were prepared by oxidation of **45** and **46** with m-chloroperbenzoic acid (mCPBA). The synthesis of the optically active oxides was conducted by the stereoselective conversion of the chiral sulfuranes using RuO_4, according to the procedure reported earlier [66].

The authors have also reported kinetic studies of the thermal racemization of optically active spirosulfuranes **45**, **46**, and the oxide **48**. Activation parameters

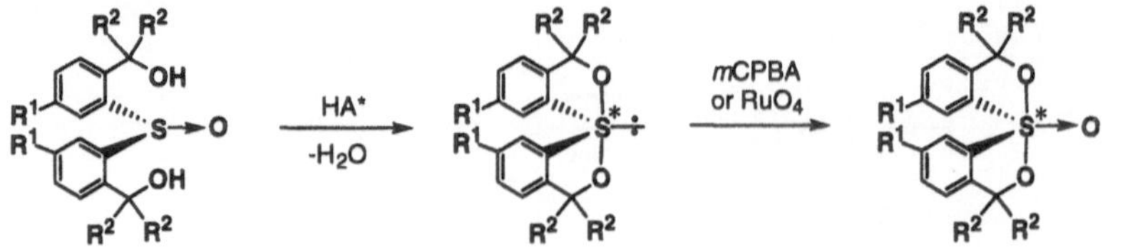

47

45: R^1 = Me, R^2 = H
$$\left(\begin{array}{l} \Delta H^{\neq} = 28.4 \text{ kcal/mol} \\ \Delta S^{\neq} = -1.5 \text{ eu} \end{array} \right)$$

46: R^1 = CF_3, R^2 = t-Bu
$$\left(\begin{array}{l} \Delta H^{\neq} = 35.6 \text{ kcal/mol} \\ \Delta S^{\neq} = 5.6 \text{ eu} \end{array} \right)$$

48: R^1 = Me, R^2 = H
$$\left(\begin{array}{l} \Delta H^{\neq} = 26 \text{ kcal/mol} \\ \Delta S^{\neq} = 6 \text{ eu} \end{array} \right)$$

49: R^1 = CF_3, R^2 = t-Bu

HA* = (+)(1S)camphor sulfonic acid, (1R,3S)-camphoric acid, (+)-(S)-mandelic acid

Scheme 29

Table 6. Activation parameters for thermal racemization of spirosulfurane **46** in different solvents

Solvent	E_a (kcal/mol)	$\Delta H^{\neq}$ (kcal/mol)	$\Delta S^{\neq}$ (eu)
Anisole	36.4	35.6	5.6
Tetrachloroethylene	35.5	34.7	4.0
Pyridine/H_2O	27.1	26.3	-14.2
$CH_3COOH/(CH_3CO)_2O$	34.4	33.6	1.1

of the racemization in anisole are shown in Scheme 29. Furthermore, the solvent effect on the rate of the racemization of the sulfurane **46** was examined, and the results are collected in Table 6.

3.2
Unsymmetrical Spiro-Oxysulfuranes [10-S-4], [10-S-5]

Recently, Kapovits *et al.* have reported three diaryldiacyloxyspirosulfuranes **50**, **51**, and **52**, each having a five-, six-, and seven-membered spiroring, respectively [67]. Their molecular structures were determined as slightly distorted trigonal bipyramidal (TBP) geometries, resulting in chiral molecular structures, by X-ray crystallographic analysis, as shown in Fig. 5. The structural parameters about sulfur in the spirosulfuranes [10-S-4(C2O2)] with aromatic carbon and acyloxy-oxygen are similar to one another, and are not greatly influenced by the shape and size of the spirorings. The individual apical S-O bond lengths can not be correlated with the size of the spirorings, but the corresponding O-S-O distances increase gradually from **50** to **52**, paralleling the decrease in O-S-O distances in diacyloxysulfuranes, which are longer than those in dialkoxy analogues, but shorter than in mixed alkoxyacyloxy derivatives.

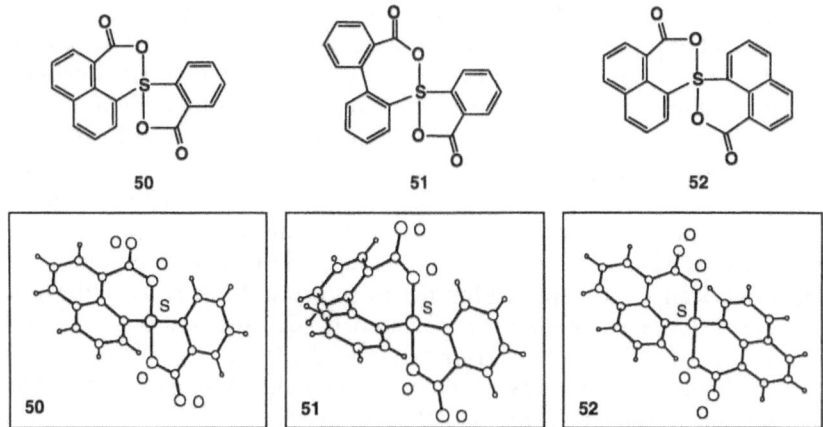

Figure 5. Structure scheme and molecular structure view of **50–52**

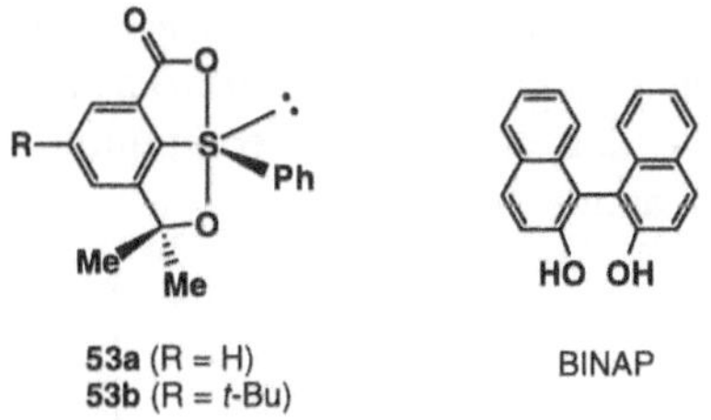

Scheme 30

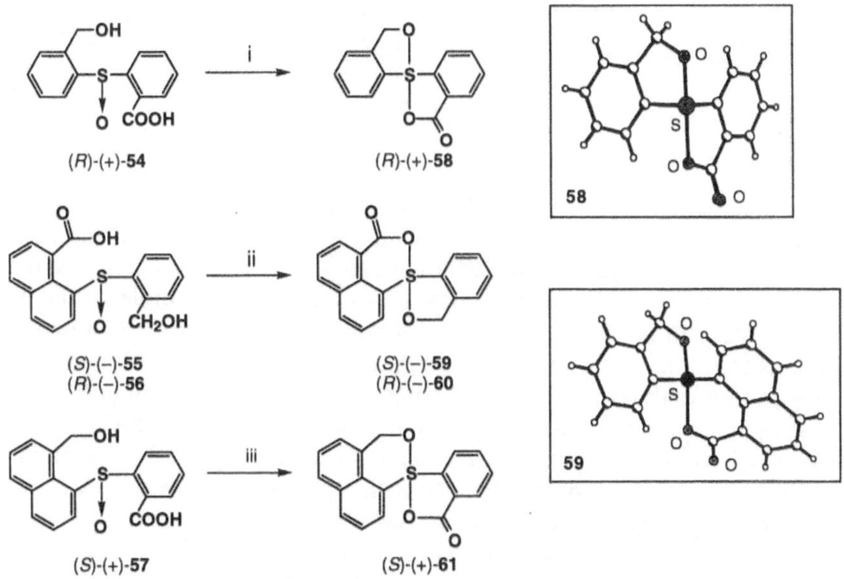

53a (R = H)
53b (R = t-Bu)

BINAP

(R)-(+)-**54** → i → (R)-(+)-**58**

58

(S)-(−)-**55**
(R)-(−)-**56** → ii → (S)-(−)-**59**
(R)-(−)-**60**

59

(S)-(+)-**57** → iii → (S)-(+)-**61**

i) AcCl, 1 equiv Et₃N, DMF, -60 °C, then 1,2-dichloroethane, 83 °C, 20 min;
ii) AcCl, 2 equiv Et₃N, DMF, -60 °C;
iii) DCC, CH₂Cl₂, 20 °C 4 h.

Scheme 31

Martin and Drabowicz have reported that the first optically active spirosulfuranes **53** containing a tridentate ligand have been obtained by a nonclassical resolution procedure with optically active 2,2-dihydroxy-1,1'-binaphthol (BINAP), as shown in Scheme 30 [68]. Compound **53a** was found to be optically stable under conditions in which the symmetrical spirosulfuranes **1** with bidentate ligands lose their optical activity rapidly.

The prochiral sulfoxide diols were dehydrated under asymmetric conditions to yield spirosulfuranes with a rather low enantiomeric excess (below 5%), as described above. Further, Szabó et al. have reported the first stereospecific synthesis of optically active spirosulfuranes as shown in Scheme 31 [69]. Optical active spirosulfuranes **54–57** were prepared by dehydration of the optically active diaryl sulfoxides **58–61**, carrying reactive CH_2OH and COOH substituents in very high yields. The molecular structures, including the absolute configurations, were determined by X-ray crystallography.

Kálmán et al. have reported the first preparation of a diaryl(acyloxy)sulfonylaminospirosulfurane derivative **62** [10–S–4(C2ON)] [70]. The sulfoxide having both sulfilimino- and carboxyl groups was treated with Ac_2O-pyridine or $(CF_3CO)_2O$-DMF at 0–20 °C to give the spirosulfurane **62** in high yield, as shown in Scheme 32. The molecular structure of compound **62** was determined by single-crystal X-ray diffraction analysis.

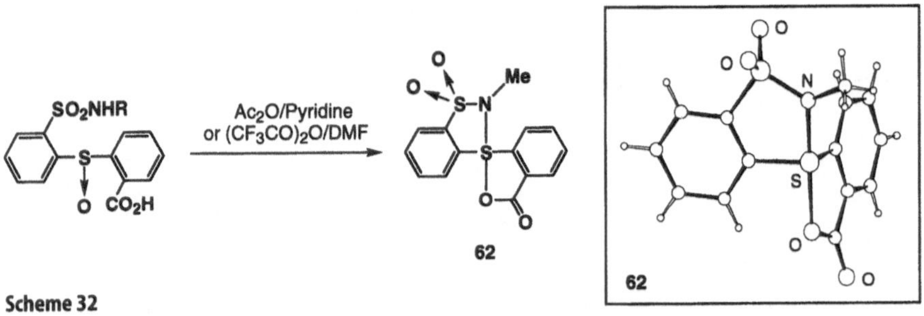

Scheme 32

The reactions of sulfur ylids with carbonyl compounds, known as the Corey-Chaykovsky reaction, [71] are well known to give oxiranes, in sharp contrast to those of phosphorus ylids, which afford olefins (the Wittig reaction). It is interesting to study whether tetracoordinate 1,2-oxaphosphetanes undergo the Wittig-type reaction giving an olefin or reductive elimination giving an oxirane. Quite recently, Kawashima and Okazaki et al. have reported new spiro-oxysulfuranes and -oxysulfurane S-oxides bearing a four-membered ring [72]. The 3-unsubstituted tetracoordinated 1,2-oxathietane **63** was synthesized as a thermally stable compound, but was not isolated because it proved too moisture-sensitive. The stable 1,2-oxathietane **64** with a phenyl group at the 3-position, however, was prepared by oxidative cyclization of **65**, as shown in Scheme 33. The molecular structure of **64** was determined by X-ray crystallography. In the thermolysis of (E)-**64**, an equilibrium between (E) and (Z) diastereoisomers was observed, and products **66** and **67** (path a) and **68**, **69**, and **70** (path b) were obtained. The thermolysis of oxathietane (E)-**64** did not give the olefin, in sharp contrast to the oxetanes containing group 13, 14, and 15 elements, [73] indicat-

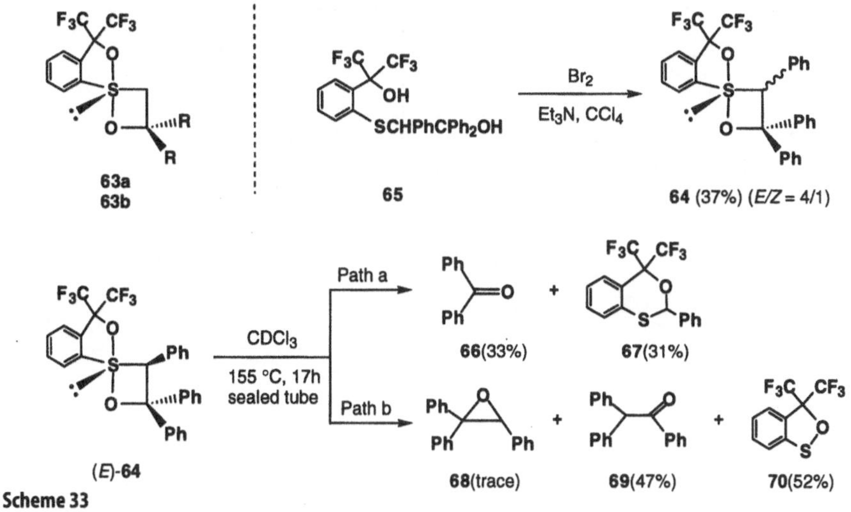

Scheme 33

ing that the bond energy of a chalcogen-oxygen double bond is not sufficient to undergo a Wittig-type reaction.

The corresponding stable sulfurane oxide 71 was prepared by the oxidation of 72 with mCPBA in the presence of Na$_2$HPO$_4$, as shown in Scheme 34 [74] and Fig. 6. The molecular structure of 71 was also determined by X-ray crystallography. In contrast, the thermolysis of 71 gave the corresponding oxirane 73 in almost quantitative yield, but the expected olefin was not obtained at all. These results suggest that 1,2-oxathietane derivatives are in fact the intermediates of Corey-Chaykovsky reaction of the sulfonium and oxosulfonium ylids.

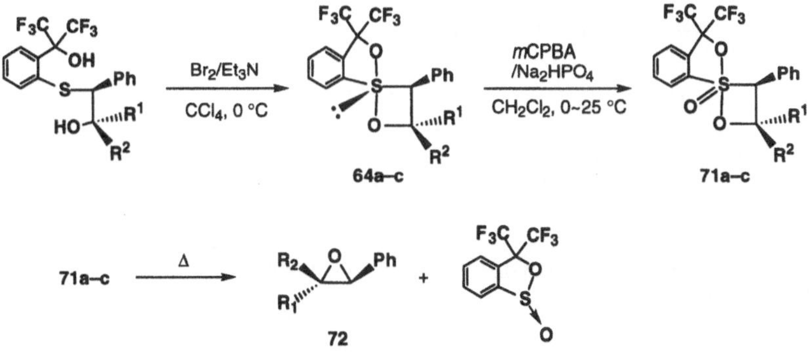

Scheme 34   a: R^1 = R^2 = CF$_3$, b: R^1 = Ph, R^2 = CF$_3$; c: R^1 = CF$_3$, R^2 = Ph

Furthermore, they have calculated energy values for the concerted oxirane formation from the 1,2-oxathietane using the model compound 74 by ab initio calculations, as shown in Fig. 7 [74b]. This result is the first location of an oxirane formation pathway from 1,2-oxathietane as a carbon-oxygen ligand coupling reaction of a sulfurane.

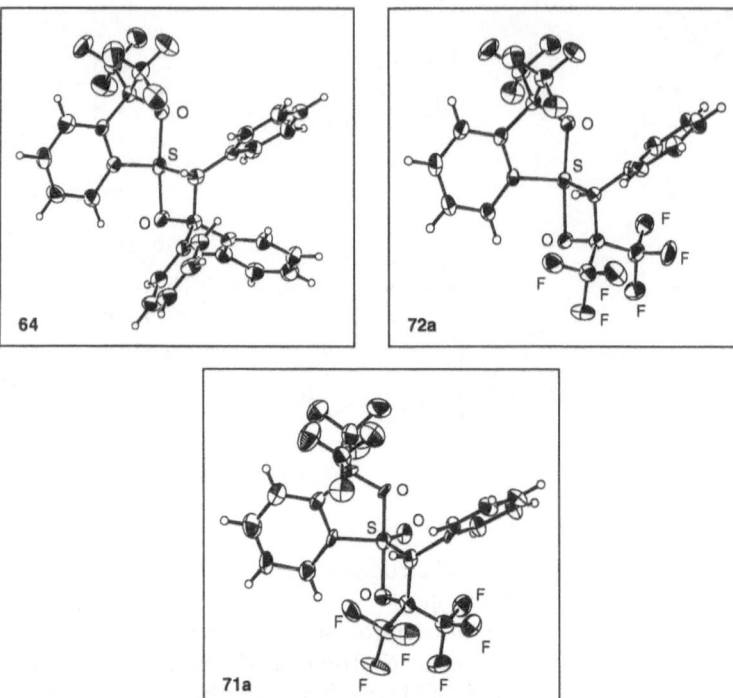

Figure 6. ORTEP views of spirosulfurane **64** and **72a**, and spirosulfurane *S*-oxide **71a**

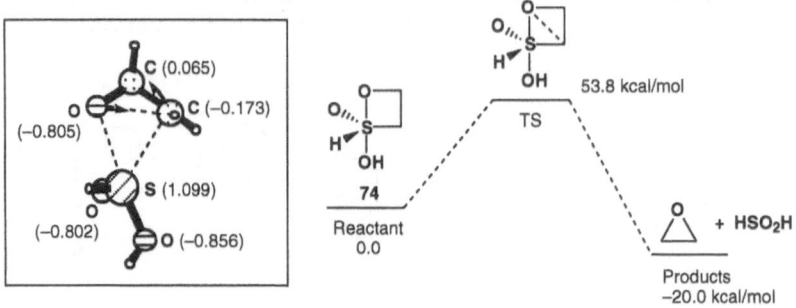

Figure 7. Structure of TS and energy diagram for the oxirane formation from model compound **74** (RHF/4–31G*)

3.3
Cationic Sulfurane [10–S–4]⁺ and Dicationic Sulfurane [10–S–4]²⁺

Akiba *et al.* have reported the cationic monoazasulfuranes [10–S–4(C2NY), Y=C, O, Cl] having an apical N-S, bond formed by transannular interaction between the amino and the sulfonio groups as shown in Fig. 8 [75]. Sulfuranes **75a,b** are the first examples of sulfuranes with an apical alkyl group. The central sulfur atom of **75a,c,e** has a distorted trigonal bypyramidal bonding geometry.

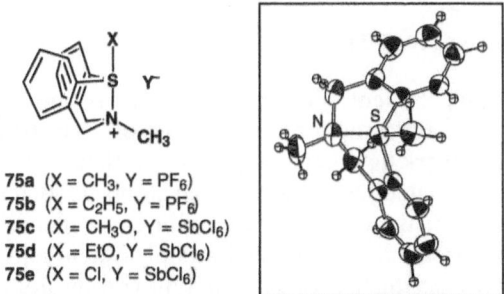

75a (X = CH₃, Y = PF₆)
75b (X = C₂H₅, Y = PF₆)
75c (X = CH₃O, Y = SbCl₆)
75d (X = EtO, Y = SbCl₆)
75e (X = Cl, Y = SbCl₆)

Figure 8. Structure scheme and ORTEP view of sulfurane **75a**

The S-N bond distances exhibited are 2.10–2.50 Å, which are within the sum of the van der Waals radii of nitrogen and sulfur (3.35 Å). A linear relationship is observed between the ^{1}H and ^{13}C NMR chemical shifts of the N-methyl group and the Hammett σ_m-substituent constants.

In the course of the studies on transannular interaction [76] between heteroatoms, Furukawa and co-workers have recently reported some interesting new stable azasulfuranes [77], sulfurane dications bearing two sulfur atoms [78], and selenurane dications bearing two sulfur and selenium atoms as ligands which are annelated in two five-membered diazocine skeletons. As shown in Scheme 35, the central sulfur atom of the tris-sulfide **76** can be converted

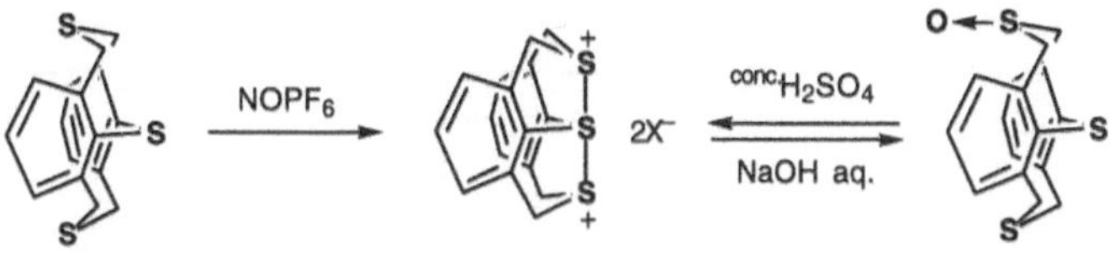

76
twin-chair

NOPF₆ →

77
twin-boat
(X = PF₆, HSO₄)

2X⁻

^{conc}H₂SO₄ ⇌ NaOH aq.

78
twin-chair

Scheme 35

into the tetracoordinated sulfurane [10–S–4(C2S2)]²⁺ **77** by employing a suitable one-electron oxidant such as NOBF₄, or by dissolving the monoxide **78** in concentrated H₂SO₄ [78]. The conformation of the tris-sulfide **76** is a chair-chair form (TC) which is transformed instantaneously into the boat-boat form (TB) on treatment with NOBF₄ due to a transannular interaction between the three sulfur atoms. Ab initio MO calculations using STO-3G$^{(*)}$ as a basis set also reveal that the most stable conformations of **76**, **77**, and **78** are the twin chair and the twin boat forms, respectively.

3.4
Tricoordinated Sulfuranes [10–S–3]

The chemistry of π-hypervalent heterocyclic systems is one of the important fields in that of hypervalent compounds. A number of π-electron systems containing a 10–S–3 framework have been prepared, and their structures and reactivities have been investigated [79]. In general, all atoms bonded to central atom in π-sulfuranes such as trithiapentalene are in the same plane. The equatorial bond has the characterization of a double bonding system and the two apical bonds are a 3-center 4-electron bond. As the sp² type carbon atom or heteroatom having a lone electron pair is located in the two five-membered ring systems, a 12π conjugated system is constructed and plays an important part in the stability of π-sulfurane.

Akiba *et al.* have reported the characterization of a sulfurane [10–S–3] intermediate during the reaction of thiadiazoline or thiazoline derivatives with activated acetylenes [80]. In order to investigate the stabilities of the 3c-4e bond in this species, sulfuranes fused with pyrimidine rings **79** were prepared and the structures were determined to be planar by X-ray crystallographic analysis, as shown in Fig. 9. The bond energy of the hypervalent N-S-N bond was estimated by using the rotation energy of the pyrimidine ring, as shown in Table 7.

Iwasaki *et al.* have reported the syntheses and structure determination of many π-sulfurane species. Recently, they isolated the 1,6,6a-trithia(6a-S^IV)pentalenes **81** and **82** by reaction pathways as shown in Scheme 36 [81]. Both molecular structures were determined by X-ray crystallography.

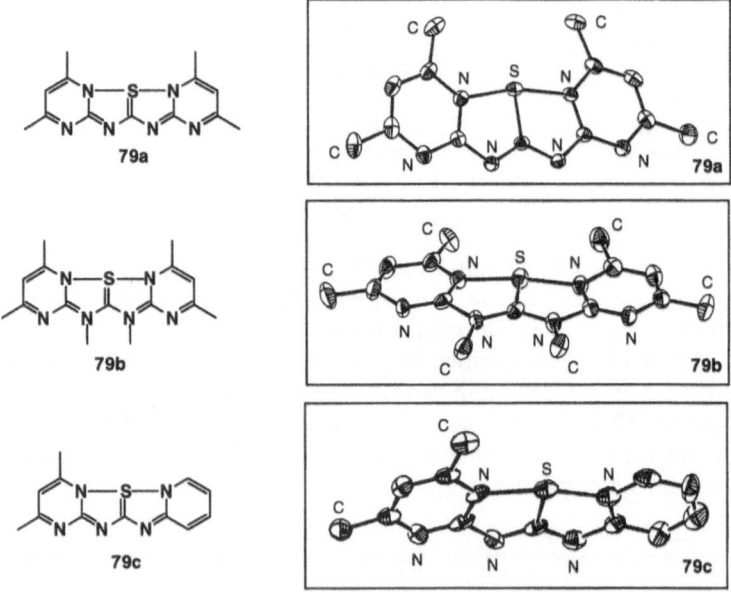

Figure 9. Structure scheme and ORTEP views of 79a–c

Table 7. Activation parameters of S-N bond cleavage of **79b** and **80** in CD_2Cl_2

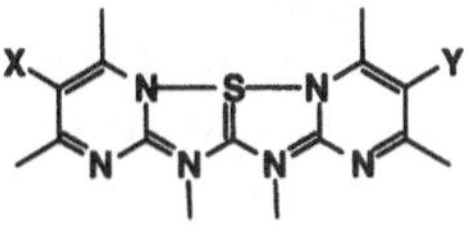

79b (X = Y = H)
80 (X = H, Y = Br)

Compound	X	Y		Tc (°C)	$\Delta G^{\neq}_{Tc}$	$\Delta G^{\neq}_{298}$	$\Delta H^{\neq}$ (kcal/mol)	$\Delta S^{\neq}$ (eu)
79b	H	H		45	16.7	16.6	15.9±1.1	-2.4±3.4
80	H	Br	H-side	75	18.2	18.3	18.8±0.3	1.8±1.0
	H	Br	Br-side	24	15.6	15.5	16.1±1.3	2.2±0.9

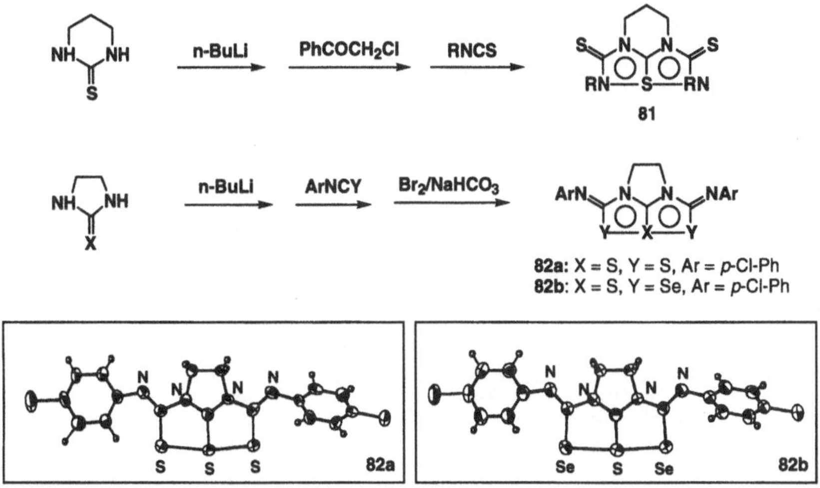
n-BuLi
PhCOCH2Cl
RNCS
81
n-BuLi
ArNCY
Br2/NaHCO3
82a: X = S, Y = S, Ar = p-Cl-Ph
82b: X = S, Y = Se, Ar = p-Cl-Ph
82a
82b

Scheme 36

They have also reported the reactivities of these π-sulfurane derivatives upon treatment with reducing agents or transition metal reagents. In the course of studying the reactivity of 6a-thiatetraazapentalene derivatives, the reaction of **81** with $NaBH_4$ gave the ring-opening compound **83** with release of the hyperva-

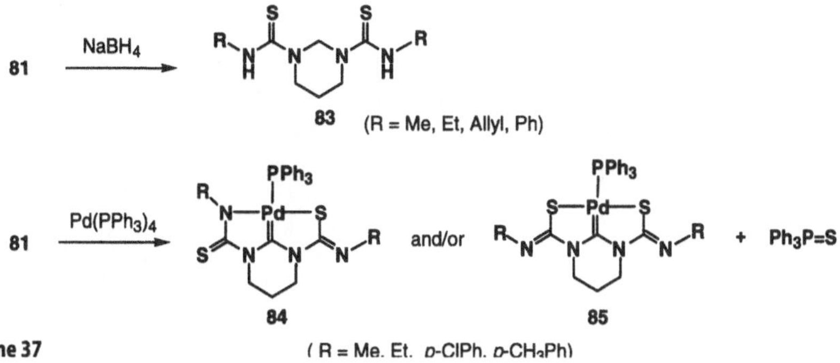
81
NaBH4
83 (R = Me, Et, Allyl, Ph)
81
Pd(PPh3)4
and/or
+ Ph3P=S
84
85
(R = Me, Et, p-ClPh, p-CH3Ph)

Scheme 37

lent sulfur as shown in Scheme 37 [82]. This result indicates that the carbon atom of the $S^{IV}=C$ bond of **83** has an electrophilic nature. Furthermore, transition metal-carbene complexes were obtained from **84** and **85** by treatment with $[Pt(PPh_3)_4]$, $[Pd(PPh_3)_4]$, and $[Rh(PPh_3)_4]$, respectively [83].

Reid *et al.* have reported that the reaction of thiadiazolo pyrimidine derivatives with isocyanates, isothiocyanates, and isoselenacyanates, with elimination of acetonitrile and concomitant addition of two molecules of the hetero-cumulene, gave the corresponding π-sulfuranes **86–93**, as shown in Fig. 10 [84]. The molecular structures of **87** (R=Et, Ph) were determined by X-ray crystallographic analysis [85].

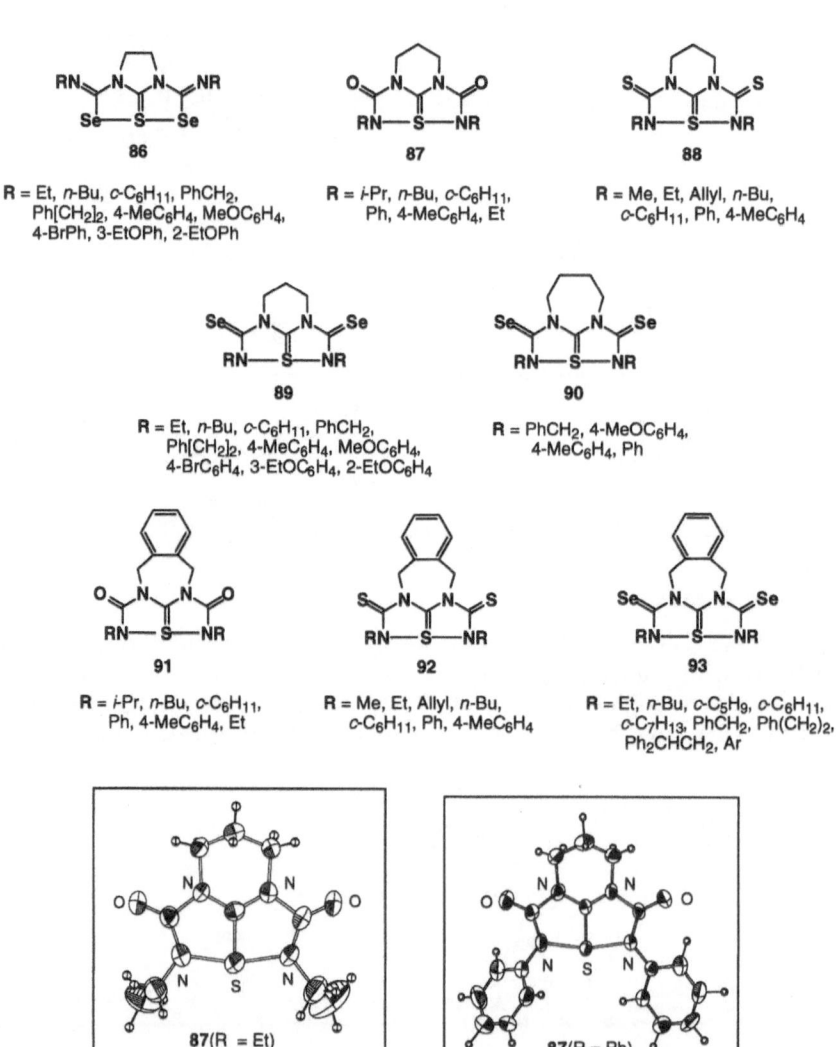

86

R = Et, *n*-Bu, *c*-C$_6$H$_{11}$, PhCH$_2$, Ph[CH$_2$]$_2$, 4-MeC$_6$H$_4$, MeOC$_6$H$_4$, 4-BrPh, 3-EtOPh, 2-EtOPh

87

R = *i*-Pr, *n*-Bu, *c*-C$_6$H$_{11}$, Ph, 4-MeC$_6$H$_4$, Et

88

R = Me, Et, Allyl, *n*-Bu, *c*-C$_6$H$_{11}$, Ph, 4-MeC$_6$H$_4$

89

R = Et, *n*-Bu, *c*-C$_6$H$_{11}$, PhCH$_2$, Ph[CH$_2$]$_2$, 4-MeC$_6$H$_4$, MeOC$_6$H$_4$, 4-BrC$_6$H$_4$, 3-EtOC$_6$H$_4$, 2-EtOC$_6$H$_4$

90

R = PhCH$_2$, 4-MeOC$_6$H$_4$, 4-MeC$_6$H$_4$, Ph

91

R = *i*-Pr, *n*-Bu, *c*-C$_6$H$_{11}$, Ph, 4-MeC$_6$H$_4$, Et

92

R = Me, Et, Allyl, *n*-Bu, *c*-C$_6$H$_{11}$, Ph, 4-MeC$_6$H$_4$

93

R = Et, *n*-Bu, *c*-C$_5$H$_9$, *c*-C$_6$H$_{11}$, *c*-C$_7$H$_{13}$, PhCH$_2$, Ph(CH$_2$)$_2$, Ph$_2$CHCH$_2$, Ar

87(R = Et)

87(R = Ph)

Figure 10. π-Sulfuranes **86–93** and molecular structure of **87**

Furukawa *et al.* have reported the first isolation and crystal structure of the sulfenium cationic species stabilized by two neighboring amonio groups, as shown in Scheme 38 [86]. The reaction of lithium substituted benzene derivative **94** with sulfur dichloride gave the corresponding sulfenium chloride **92**, having Cl^- as the counter anion. The sulfenium salt **95** was treated with potassium hexafluorophosphate (KPF_6) to give the corresponding sulfenium salt **96**, having PF_6^- as a counteranion.

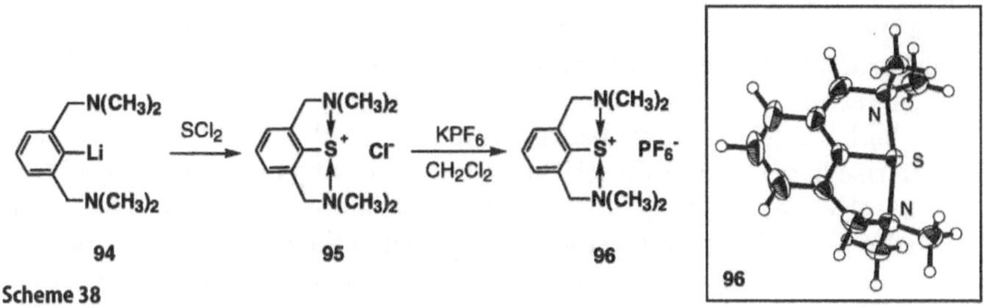

Scheme 38

3.5
Persulfonium Salts [10–S–5]⁺ and Persulfuranes [12–S–6]

The first organopersulfonium salts [10–S–5(C2O2F)]⁺, [10–S–5(CO4)]⁺ and persulfuranes [12–S–6(C2O2F2)], [12–S–6(C2O4)] were synthesized by Martin et al. [87].

Recently, Furukawa *et al.* have reported the stable penta- and hexacoordinate hypervalent sulfur compounds, [10–S–5(C2N2Cl)] **98** and [12–S–6(C2N2Cl2)] **97**, which are stabilized by the transannular interaction between nitrogen and sulfur atoms, as shown in Scheme 39 [88].

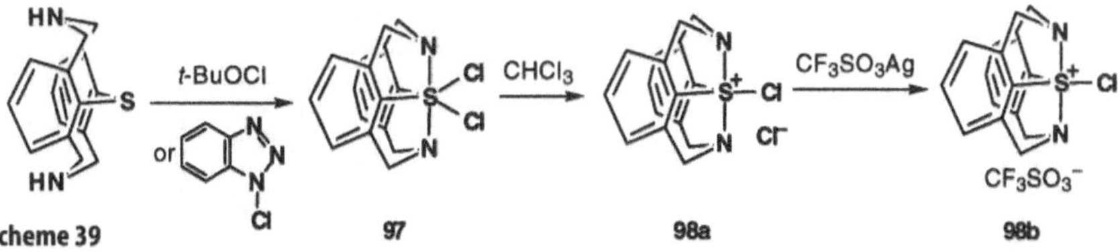

Scheme 39

4
Organosulfuranes [10–S–4(C4)] Having Only Carbon Ligands

As described in the previous section, several stable organic hypervalent compounds having group 16 elements as a central atom have been reported. Most of these compounds have strong electronegative ligands (oxygen, nitrogen, and halogen) such as the 1,1,1,3,3,3-hexafluorocumyloxy and 2-acyloxyaryl ligands found by Martin and Arhart [14] and Kapovits and Kalman [15]. In contrast, those having only carbon ligands [10–S–4(C4)] were long believed to be unstable and hence are less well known. The formation of tetraphenyl sulfurane was first suggested by Wittig and Fritz in 1952 [26], and numerous works have since been reported on the formation of tetraaryl sulfuranes. Among hypervalent sulfur compounds bearing four carbon ligands, only tetra(pentafluorophenyl)sulfurane has been observed as an intermediate, by a ^{19}F NMR experiment at low temperature [89].

Thus, sulfurane species [10–S–4(C4)] have been considered to be intermediates or transition states in the reactions of the corresponding onium salts or oxides with various organometallic reagents. Especially in the case of sulfurane species [10–S–4(C4)], no one has succeeded in direct detection except for one example. Therefore, little remains known about the properties of these compounds.

On the basis of the above background, Sato et al. tried to synthesize and characterize hypervalent sulfur(IV) compounds having only carbon ligands in order to solve the remaining problems for the isolation of sulfurane(IV), having four carbon ligands. Several factors are required in the carbon ligands, namely, four simple aryl groups should be used and five-membered annelation such as biphenylylene should be used. They attempted the preparation of the nine target compounds involving Se and Te analogues, as shown in Fig. 11 [90].

For these compounds, the direct detection of unstable species was conducted by NMR experiments at low temperature, while stable and metastable compounds were isolated and their structures were determined by X-ray crystallo-

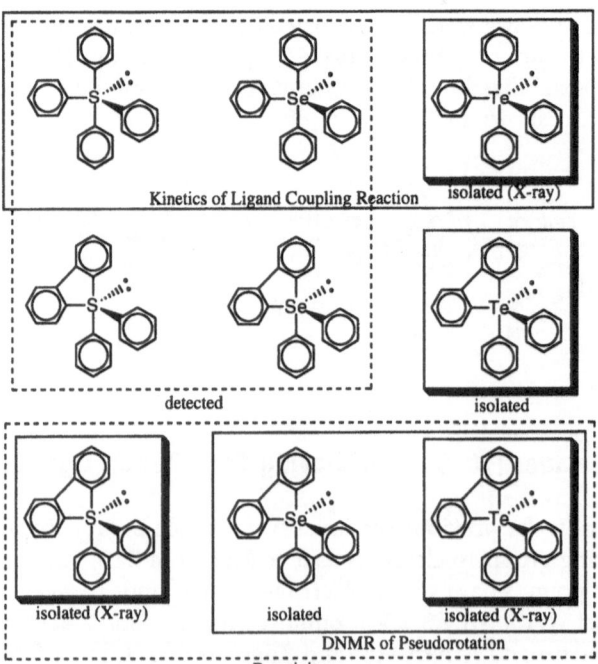

Figure 11. The target compounds of tetraarylchalcogenurane(IV)

i, 1.0 eq. PhLi, THF, -78 °C;
ii, 0.2 M aq. HPF$_6$ (Y=PF$_6$);
Scheme 40   iii, 1) 1.0eq. CF$_3$SO$_3$Si(CH$_3$)$_3$, THF, -78 °C, 2) 2.0 eq. PhLi, -78 °C.

graphic analysis. Furthermore, theoretical studies of sulfuranes having only carbon ligands were reported.

Tetraphenylsulfurane is the simplest but most unstable among the tetraaryl-sulfurane species. Although only tetraphenyltellurane has been isolated, and its crystal structure has been determined, tetraphenylsulfurane and -selenurane have neither been detected directly nor isolated in the earlier studies. Furukawa and Sato reinvestigated the reactions of triphenyl sulfonium salt **99** and diphenyl sulfoxide **100** with phenyllithium (PhLi) in tetrahydrofuran (THF) in order to attempt detection of tetraphenylsulfurane [10–S–4(C4)] **101** directly by ^{1}H, ^{13}C, and CH-COSY NMR experiments at low temperature, as shown in Scheme 40 [91]. As results, the detection of tetraphenyl sulfurane has been achieved by low temperature NMR studies while tetraphenyl tellurane has been isolated [92].

These NMR experiments at low temperature and the product analysis indicate that both triphenylsulfonium salt and diphenyl sulfoxide react with PhLi to give an identical tetraphenyl sulfurane **101** as a discrete intermediate at low temperature, which, on warming to room temperature, decomposes to diphenyl sulfide **102** and biphenyl **103**. Interestingly, the four phenyl groups in the sulfurane **101** become spectroscopically identical, suggesting that the pseudorotation takes place rapidly even at low temperature such as –105 °C.

One of the prominent reactions observed for these species is the ligand coupling reaction [93] of the hypervalent compounds incipiently formed in the reactions of triphenyl chalcogenonium salts with aryllithium [1d,e,j]. The mechanism for the reaction is considered to take place concertedly between the two ligands at the central atom of chalcogenuranes. However, no kinetic studies on the coupling reactions of the unstable hypervalent compounds have been reported, and it is still unclear whether this reaction proceeds through a concerted unimolecular pathway or involves other multi-step processes. Furukawa and Sato carried out kinetic studies of the ligand coupling reactions for these hypervalent chalcogen compounds in order to determine the order of thermal stability among the chalcogenuranes using tetraphenyl derivatives. The results on the kinetic studies for the ligand coupling reaction of tetraphenyl sulfurane **101** using a variable temperature NMR technique are shown in Table 8 [94].

Table 8. Activation parameters of ligand coupling reaction of tetraphenyl sulfurane 101[a]

Compound (solvent)	k_1 (s^{-1})		E_{act} (kcal/mol)	$\Delta G^{\neq}_{298}$ (kcal/mol)	$\Delta H^{\neq}$ (kcal/mol)	$\Delta S^{\neq}$ (eu)
Ph_4S (101) (THF-d_8)	2.48×10^{-4} ($-67\,°C$)	1.22×10^{-4} ($-72\,°C$)	10.9	17.5	10.5	-23.5
	5.00×10^{-5} ($-77\,°C$)	3.25×10^{-5} ($-82\,°C$)				

[a] Values shown are least-square treatments of Arrhenius and Eyring plots.

The 2,2'-biphenylylene diphenyl sulfurane [10–S–4(C4)] has already been postulated by Trost et al. [38] and Hori et al. as an intermediate in the reaction of 2,2'-biphenylylene phenyl sulfonium salt with phenyl lithium. However, they could not directly prove the existence of the corresponding sulfurane as an intermediate. Sato and Furukawa tried to obtain the first crucial evidence for formation of 2,2'-biphenylylene diphenyl sulfurane **105** in the reactions of 2,2'-biphenylylene phenyl sulfonium salt **104** with PhLi in a THF solution by low temperature ^{1}H, ^{13}C and CH-COSY NMR experiments [95]. By elevating the temperature of a THF solution to room temperature, each solution became colorless, to give phenyl o-terphenyl sulfide **106** quantitatively, after work up, as shown in Scheme 41.

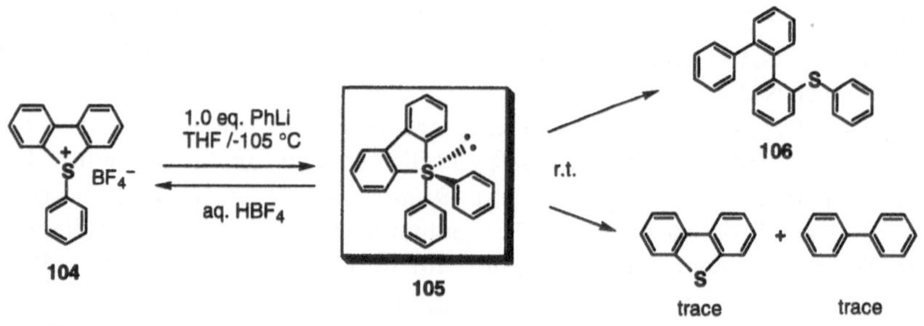

Scheme 41

Earlier, bis(2,2'-biphenylylene)sulfurane **106** was postulated by Trost and LaRochelle [38b] but to date its synthesis has not been successful. Recently, the first synthesis and structural characterization of a new stable sulfurane with four carbon-sulfur bonds, bis(2,2'-biphenylylene)sulfurane **106** have been successfully performed, as shown in Scheme 42 [29]. The structure of **106** was confirmed by X-ray diffraction analysis.

Compound **106** in the solid-state has a slightly distorted pseudo-trigonal-bipyramidal (Ψ-TBP) geometry with two apical S-C bonds, two equatorial S-C bonds, and the lone-pair electrons in the third equatorial position. The results represent the first structural characterization of an organosulfurane containing two different kinds of S-C bonds.

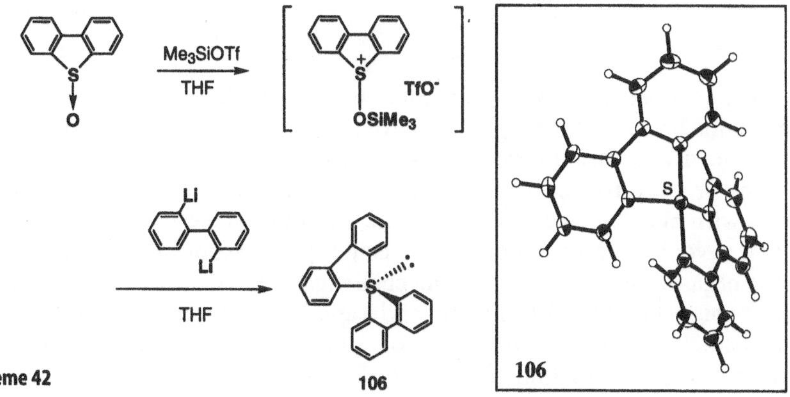

Scheme 42   **106**   **106**

The structure of **106** in THF solution was characterized by NMR spectroscopy. From the ^{1}H and ^{13}C NMR spectra, each aryl ring of the biphenylylene was magnetically equivalent at room temperature. Furthermore, these signals were unchanged at –100 °C, suggesting that Berry pseudorotation must take place rapidly on the NMR time scale even at low temperature.

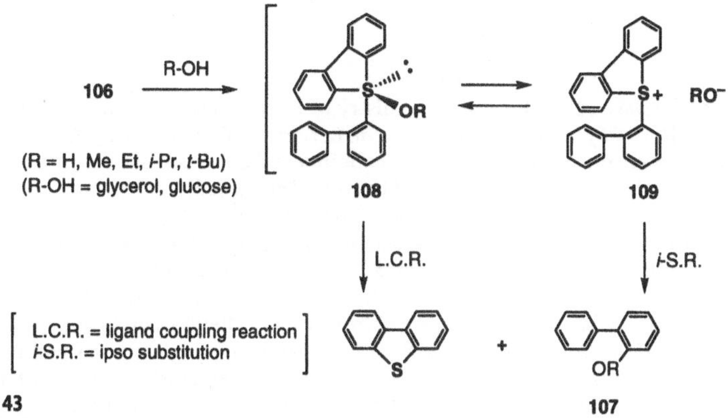

Scheme 43   **107**

Sulfurane **106** readily reacts with various alcohols and phenols to give the corresponding 2-alkoxy or 2-phenoxybiphenyls **107** in high yields through the formation of oxy-sulfuranes [10-S-4(C3O)] **108** or sulfonium salts **109**, as shown in Scheme 43 [96]. Not only mono-, but also di-, tri-, and even polyols such as glycerol and even glucose react with sulfurane **106** to give the respective polyol polybiphenyl ethers. The intermediate sulfurane **108** or sulfonium salt **109** is unstable, and has not been isolated. The authors carried out a reaction with phenol having a nitro group as an electron-withdrawing group, and succeeded in the isolation of the corresponding intermediate at room temperature [97].

5
Conclusion

This chapter does not cover the whole literature on hypervalent sulfur compounds; this has been thoroughly reviewed by Martin *et al.* and by Mikolajczyk and Drabowicz. Before the mid-1970s, sulfuranes were considered to be intermediates or only transition states in many reactions of organosulfur compounds with nucleophiles. Their existence, except for a few compounds such as SF_4, had never been proven nor ever detected spectroscopically, although many famous organic chemists took up the challenge of the detection of sulfuranes. Since the pioneering work by Martin and Kapovits, many stable sulfuranes have been prepared and the chemistry of sulfuranes has helped trigger the development of heteroatom chemistry. The authors have been interested in high-coordinated sulfur compounds for a long time and, recently, have succeeded in the detection of tetraphenyl sulfuranes by low temperature NMR techniques, and finally in the isolation of bis(2,2'-biphenylylene)sulfurane as a stable sulfurane, which had long been sought. The isolation of stable sulfuranes has stimulated many organic chemists to undertake the development of heteroatom chemistry as a new field. The authors would like to contribute to further aspects of this chemistry, not only of sulfur but also of other chalcogen atoms.

References

1. For some reviews and books of the chemistry of hypervalent compounds, see: (a) Hayes RA, Martin JC (1985) Sulfurane chemistry. In: Bernardi F, Csizmadia IG, Mangini A (eds) Organic sulfur chemistry; theoretical and experimental advances. Elsevier, Amsterdam, p 408; (b) Bergman J, Engman L, Sieden J (1986) In: Patai S (ed) The chemistry of organic selenium and tellurium compounds. Wiley, Chichester, chap 14; (c) Rauchfuss TB (1987) Acidity, hydrogen bonding and self-association in organic and organometallic compounds of selenium and tellurium. In: Patai S (ed) The chemistry of organic selenium and tellurium compounds, vol 2. Wiley, Chichester, chap 6, p 339; (d) Oae S (1988) Ligand coupling reactions within hypervalent species. In: Oae S (ed) Reviews on heteroatom chemistry, vol 1. MYU, Tokyo, p 304; (e) Oae S (1991) Ligand coupling reactions within hypervalent species I. On sulfur and related atoms of group VI elements. In: Oae S (ed) Reviews on heteroatom chemistry, vol 4. MYU, Tokyo, p 195; (f) Oae S, Doi JT (1991) Organic sulfur chemistry: structure and mechanism. CRC Press, Boca Raton, p 433; (g) Furukawa N (1990) Reactions of sulfoxides with organometallic reagents: is sulfurane an intermediate? In: Block E (ed) Heteroatom chemistry. VCH, New York, p 165; (h) Drabowicz J, Lyzwa P, Mikolajczyk M (1993) High-coordinated sulfur compounds. In: Patai S, Rappoport Z (eds) The chemistry of sulphur-containing functional groups. Wiley, New York, p 799; (i) Furukawa N, Sato S (1996) Mononuclear hydrocarbons carrying nuclear substituents containing selenium or tellurium. In: Sainsbury M (ed) Rodd's chemistry of carbon compounds, 2nd edn, vol III. Elsevier Science, Amsterdam, p 469; (j) Finet JP (1998) Ligand coupling reactions with heteroatomic compounds. Elsevier, Oxford; (k) Akiba K-y (1999) Chemistry of hypervalent compounds. VCH, New York
2. March J (1985) Advanced organic chemistry. Wiley, New York
3. Pauling L (1960) The nature of chemical bond. Cornell University Press, New York
4. (a) Michaelis A, Schifferdecker O (1873) Chem Ber 6:993; (b) Ruff O, Heizelmann A (1911) Z Anorg Allg Chem 72:63

5. Innes EA, Csizmadia IG, Kanada Y (1989) J Mol Structure (Theochem) 186:1
6. Robinson EA (1989) J Mol Struct (Theochem) 186:9
7. (a) Block E (1990) Heteroatom chemistry. VCH, New York; (b) McEwen WE (1990) Hetero-atom Chem, VCH, International journal since (c) 1st International Symposium on Hetero-atom Chemistry was organized by Professor S. Oae at Kobe, Japan in 1987
8. (a) Hatch RJ, Rundle RE (1951) J Am Chem Soc 73:4321; (b) Rundle RE (1963) J Am Chem Soc 85:112; (c) Musher JI (1969) Angew Chem Int Ed Engl 8:54
9. (a) Chen MML, Hoffmann R (1976) J Am Chem Soc 98:1647; (b) Reed AE, Schleyer P von R (1990) J Am Chem Soc 112:1434; (c) Kutzelnigg W (1984) Angew Chem Int Ed Engl 23:272
10. (a) Reed AE, Weinhold F (1986) J Am Chem Soc 108:3586; (b) Minoura M, Tamagaki T, Akiba K-y, Modrahowski C, Sudau A, Seppelt K, Walllenbauer S (1996) Angew Chem Int Ed Engl 35:2660
11. Sato S, Yamashita T, Takahashi O, Furukawa N, Yokoyama M, Yamaguchi K (1997) Tetra-hedron 53:12,183
12. (a) Ahmed L, Morrison JA (1990) J Am Chem Soc 112:7411; (b) Sato S, Yamashita T, Horn E, Furukawa N (1996) 15:3256; (c) Minoura M, Sagami T, Miyasato M, Akiba K-y (1997) Tetrahedron 53:12,195; (d) Sato S, Ueminami T, Horn E, Furukawa N (1997) J Organomet Chem 543:77; (e) Sato S, Arakawa H, Horn E, Furukawa N (1998) Chem Lett 213
13. (a) Johnson CR, McCants D (1965) J Am Chem Soc 87:5404; (b) Kice J K (1980) Adv Phys Org Chem 17:65
14. Martin JC, Arhart R (1971) J Am Chem Soc 93:2339
15. Kapovits I, Kalman A (1971) J Chem Soc Chem Commun 649
16. Mutterties EL, Schunn RA (1966) Quat Rev 20:245
17. (a) Oae S, Uchida Y (1991) Acc Chem Res 24:202; (b) Oae S (1989) J Mol Struct (Theochem) 186:321
18. (a) Ugi I, Marquarding D, Klusacek H, Gillespie P, Ramirez F (1971) Acc Chem Res 4:288; (b) Berry RS (1960) J Chem Phys 32:933
19. (a) Holmes RR (1979) Acc Chem Res 12:257; (b) Luckenbach R (1973) Dynamic stereo-chemistry of the pentacoordinated phosphorus and related elements. George Thieme, Stuttgart
20. Oae S, Doi JT (1991) Substitution. In: Oae S, Doi JT (eds) Organic sulfur chemistry: struc-ture and mechanism. CRC Press, Boston, chap 4, p 119
21. (a) Pross A, Shaik SS (1983) Acc Chem Res 16:363; (b) Sini G, Ohanessian G, Hiberty PC, Shaik SS (1990) J Am Chem Soc 112:1407
22. Paulmier C (1986) Selenium reagents and intermediates in organic synthesis. Pergamon, Oxford
23. Tolles MW, Gwinn WD (1962) J Chem Phys 36:119
24. Baezinger NC, Buckles RE, Maner RJ, Simpson TD (1969) J Am Chem Soc 91:5749
25. Franzen V, Mertz C (1961) Justus Liebigs Ann Chem 643:24
26. Wittig G, Fritz H (1952) Justus Liebigs Ann Chem 577:39
27. (a) Hellwinkel D (1972) Ann New York Acad Sci 192:158; (b) Hellwinkel D, Fahrbach G (1968) Justus Liebigs Ann Chem 712:1
28. Hellwinkel D, Fahrbach G (1968) Justus Liebigs Ann Chem 715:68
29. Ogawa S, Matsunaga Y, Sato S, Iida I, Furukawa N (1992) J Chem Soc Chem Commun 1141
30. Ogawa S, Sato S, Erata T, Furukawa N (1991) Tetrahedron Lett 32:3179
31. (a) Kice JL (1972) Prog Org Chem 17:147; (b) Parker AJ, Kharasch N (1959) Chem Rev 59:583; (c) Kice JL (1968) Acc Chem Res 1:58; (d) Oae S (1977) Organic chemistry of sul-fur. Plenum Press, New York
32. (a) Johnson CR, Rigau JJ (1969) J Am Chem Soc 91:5398; (b) Calo V, Scorrano G, Modena G (1969) J Org Chem 24:2020; (c) Owsley DC, Helmkamp GK, Rettig F (1969) J Am Chem Soc 91:5239
33. (a) Young PR, Huang HC (1987) J Am Chem Soc 109:1805; (b) Young PR, Ruckberg BP (1989) J Mol Struct (Theochem) 186:85

34. (a) Okuyama T, Lee JP, Ohnishi K (1994) J Am Chem Soc 116:6480; (b) Okuyama TO, Fueno T (1990) Bull Chem Soc Jpn 63:1316
35. (a) Lockchard JP, Schroeck CW, Johnson CR (1973) Synthesis 485; (b) Durst T, Lebell MJ, vanden Elzen R, Tin K-C (1974) Can J Chem 52:761; (c) Hojo M, Masuda R, Sakai Y, Fujimori K, Tsutsumi S (1977) Synthesis 789; (d) Johnson CR, McCants D (1965) J Am Chem Soc 87:5404
36. Harrington D, Weston J, Jacobus J, Mislow K (1972) J Chem Soc Chem Commun 1079
37. (a) Andersen KK, Caret RL, K-Nielsen I (1974) J Am Chem Soc 96:8026; (b) Ackerman BK, Andersen KK, K-Nielsen I, Peynircioglu NB, Yeager YA (1974) J Org Chem 39:964
38. (a) Trost BM, Schinski WL, Mantz JD (1969) J Am Chem Soc 91:4320; (b) LaRochelle RW, Trost BM (1971) J Am Chem Soc 93:6077; (c) Trost BM, Siman SD (1971) J Am Chem Soc 93:3825; (d) Trost BM, Atkins RC, Hoffman L (1973) J Am Chem Soc 95:1285
39. (a) Gilman H, Swayampatai DR (1955) J Am Chem Soc 77:337; (b) Gilman H, Swayampatai DR (1957) J Am Chem Soc 79:20
40. Kimura T, Horie Y, Ogawa S, Furukawa N, Iwasaki F (1993) Heteroatom Chem 4:243
41. Wildi BS, Taylor SW, Potratz HA (1951) J Am Chem Soc 73:1965
42. Chung S-K, Sakamoto S (1981) J Org Chem 46:4590
43. Furukawa N, Ogawa S, Matsumura K, Fujihara H (1991) J Org Chem 56:6341
44. (a) Field L (1977) Organic chemistry of sulfur. In: Oae S (ed) Plenum Press, New York, chap 6; (b) Benson SW (1978) Chem Rev 78:23
45. Ogawa S, Tajiri Y, Furukawa N (unpublished result)
46. Sato S, Furukawa N (unpublished result)
47. (a) Oae S, Furukawa N (1990) Adv Heterocycl Chem 48:1
48. Moc J, Drigo AE, Morokuma K (1993) Chem Phys Lett 204:65
49. (a) Oae S, Kawai T, Furukawa N (1984) Tetrahedron Lett 25:2569: (b) Oae S, Kawai T, Furukawa N, Iwasaki F (1987) J Chem Soc Perkin Trans 2 405
50. Kawai T, Kodera Y, Furukawa N, Oae S, Ishida M, Wakabayashi S (1987) Phosphorus Sulfur 34:139
51. Oae S, Takeda T, Wakabayashi S (1988) Tetrahedron Lett 29:4445
52. Wakabayashi S, Ishida M, Takeda T, Oae S (1988) Tetrahedon Lett 29:4441
53. Neef G, Eder U, Seeger A (1980) Tetrahedron Lett 21:903
54. (a) Teobald PG, Okamura WH (1987) Tetrahedron Lett 28:6565; (b) Teobald PG, Okamura W H (1990) J Org Chem 55:741
55. Oae S, Kawai T, Furukawa N (1984) Tetrahedron Lett 25:2549
56. Shibutani T, Fujihara H, Furukawa N, Oae S (1991) Heteroatom Chem 2:521
57. Furukawa N, Shibutani T, Matsumura K, Fujihara H, Oae S (1986) Tetrahedron Lett 27:3899
58. Furukawa N, Shibutani T, Fujihara H (1987) Tetrahedron Lett 28:5845
59. (a) Shibutani T, Fujihara H, Furukawa N (1991) Tetrahedron Lett 32:2943; (b) Furukawa N, Shibutani T, Fujihara H (1989) Tetrahedron Lett 30:7091
60. Hornbuckle SF, Livant P, Webb TR (1995) J Org Chem 60:4153
61. Furukawa N, Ogawa S, Matsumura K, Shibutani T, Fujihara H (1990) Chem Lett 979
62. Ohkata K, Ohnishi M, Yoshinaga K, Akiba K-y, Rongione JC, Martin JC (1991) J Am Chem Soc 113:9270
63. Rongione JC, Martin JC (1990) J Am Chem Soc 112:1637
64. Allenmark S, Claeson S (1993) Tetrahedron: Asymmetry 4:2329
65. Drabowicz J, Martin JC (1993) Pure & Appl Chem 68:951
66. Perozzi EF, Martin JC (1972) J Am Chem Soc 94:5519
67. Kapovits I, Rabai J, Szabo D, Czako K, Kucsman Á, Argay G, Fülöp V, Kálmán A, Koritsánszky T, Párkányi L (1993) J Chem Soc Perkin Trans 2:847
68. Drabowicz J, Martin JC (1993) Tetrahedron: Asymmetry 4:2329
69. (a) Szabó D, Szendeffy S, Kapovits I, Kucsman Á, Czugler M, Kálmán A, Nagy P (1997) Tetrahedron: Asymmetry 8:2411; (b) Vass E, Ruff F, Kapovits I, Rábai J, Szabó D (1993) J Chem Soc Perkin Trans 2:855
70. Rábai J, Kapovits I, Argay G, Koritsánszky T, Kálman A (1995) J Chem Soc Chem Commun 1069

71. Aubé J (1991) Selectivity, strategy, and efficiency in modern synthetic chemistry. In: Trost BM, Fleming I (eds) Comprehensive organic synthesis, vol 1. Pergamon, Oxford, p 822
72. (a) Kawashima T, Ohno F, Okazaki R (1994) Angew Chem Int Ed Engl 33:2094; (b) Kawashima T, Ohno F, Okazaki R (1994) Phosphorus Sulfur and Silicon 95/96:439
73. Kawashima T, Okazaki R (1996) Synlett 600
74. (a) Ohno F, Kawashima T, Okazaki R (1996) J Am Chem Soc 118:697; (b) Kawashima T, Ohno F, Okazaki R, Ikeda H, Inagaki S (1996) J Am Chem Soc 118:12,455
75. (a) Akiba K-y, Takee K, Ohkata K (1983) J Am Chem Soc 105:6965; (b) Akiba K-y, Takee K, Shimizu Y, Ohkata K (1986) J Am Chem Soc 108:6320
76. Fujihara H, Furukawa F (1989) J Mol Struct (Theochem) 186:261
77. Fujihara H, Oi N, Erata T, Furukawa N (1990) Tetrahedron Lett 31:1019
78. Fujihara H, Chiu J-J, Furukawa N (1988) J Am Chem Soc 110:1280
79. (a) Akiba K-y, Inamoto N (1980) Kagaku No Ryoiki Zokan 124:93; (b) Akiba K-y (1987) Nippon Kagaku Kaishi 1130
80. (a) Akiba K-y, Ohsugi M, Iwasaki H, Ohkata K (1988) J Am Chem Soc 110:5576; (b) Ohkata K, Ohsugi M, Kuwaki T, Yamamoto K, Akiba K-y (1990) Tetrahedron Lett 31:1603; (c) Ohkata K, Ohyama Y, Akiba K-y (1994) Heterocycles 37:859; (d) Ohkata K, Ohsugi M, Yamamoto K, Ohsawa M, Akiba K-y (1996) J Am Chem Soc 118:6355
81. (a) Yasui M, Iwasaki F (1995) J Heterocyclic Chem 32:1269; (b) Iwasaki F, Manabe N, Nishiyama H, Takada K, Yasui M, Kusamiya M, Matsumura N (1997) Bull Chem Soc Jpn 70:1267
82. Matsumura N, Tomura M, Mori O, Ukata M, Yoneda S (1987) Heterocycles 26:3097
83. (a) Matsumura N, Kawano J-i, Fukunishi N, Inoue H, Yasui M, Iwasaki F (1995) J Am Chem Soc 117:3623; (b) Yasui M, Yoshida S, Kakuma S, Shimamoto S, Matsumura N, Iwasaki F (1996) Bull Chem Soc Jpn 69:2739; (c) Iwasaki F, Manabe N, Yasui M, Matsumura N, Kamiya N, Iwasaki H (1996) Bull Chem Soc Jpn 69:2749; (d) Iwasaki F, Yasui M, Yoshida S, Nishiyama H, Shimamoto S, Matsumura N (1996) Bull Chem Soc Jpn 69:2759; (e) Manabe N, Yasui M, Nishiyama H, Shimamoto S, Matsumura N, Iwasaki F (1996) Bull Chem Soc Jpn 69:2771; (f) Iwasaki F, Nishiyama H, Yasui M, Kusamiya M, Matsumura N (1997) Bull Chem Soc Jpn 70:1277
84. (a) Lai L-L, Reid DH, Nicol RH, Rhodes JB (1994) Heteroatom Chem 5:149; (b) Lai L-L, Reid DH (1996) Heteroatom Chem 7:97; (c) Lai L-L, Reid DH (1997) Heteroatom Chem 8:13
85. Iwasaki F, Murakami H, Yamazaki N, Yasui M, Tomura M, Matsumura N (1991) Acta Cryst C46:998
86. Kobayashi K, Sato S, Horn E, Furukawa N (1998) Tetrahedron Lett 39:2593
87. (a) Michalak RS, Martin JC (1980) J Am Chem Soc 102:5921; (b)Lam WY, Deusler EN, Martin JC (1981) J Am Chem Soc 103:127; (c) Michalak RS, Martin JC (1982) J Am Chem Soc 104:1683; (d) Perkins CW, Martin JC (1983) J Am Chem Soc 105:1377
88. Fujihara H, Oi N, Erata T, Furukawa N (1990) Tetrahedron Lett 31:1019
89. (a) Sheppard WA (1971) J Am Chem Soc 93:6077; (b) Sheppard WA, Foster SS (1972/1973) J Fluorine Chem Soc 2:53
90. Sato S, Takahashi O, Furukawa N (1999) Coord Chem Rev Main Group Chem (1998) 176:483
91. Ogawa S, Matsunaga Y, Sato S, Erata T, Furukawa N (1992) Tetrahedron Lett 33:93
92. Smith CS, Lee J-S, Titus DD, Ziolo RF (1980) Organometallics 1:1338
93. (a) Barton DHR, Glover SA, Ley SV (1977) J Chem Soc Chem Commun 266; (b) Glover SA (1980) J Chem Soc Perkin Trans 1 1338; (c) Sato S, Kondo N, Furukawa N (1994) Organometallics 13:3393; (d) Sato S, Kondo N, Furukawa N (1995) Organometallics 14:5393
94. Ogawa S, Sato S, Furukawa N (1992) Tetrahedron Lett 33:7925
95. Sato S, Furukawa N (1995) Tetrahedron Lett 36:2803
96. Furukawa N, Matsunaga Y, Sato S (1993) Synlett 655
97. Sato S, Matsunaga Y, Kitagawa M, Furukawa N (1994) Phosphorus Sulfur and Silicon 95/96:447

Chemistry of Thiophene 1,1-Dioxides

Juzo Nakayama · Yoshiaki Sugihara

Department of Chemistry, Faculty of Science, Saitama University, Urawa, Saitama 338–8570, Japan. *E-mail: nakaj@post.saitama-u.ac.jp*

Thiophene oxides, where two pairs of lone pair electrons are consumed for bond formation with oxygen atoms, are no longer aromatic. Their chemistry has now grown into an important branch both in heteroatom and heterocyclic chemistry from synthetic, mechanistic, structural, and theoretical points of views. Thiophene 1,1-dioxides are most commonly prepared by oxidation of thiophenes. As unsaturated sulfones, they serve as dienophiles, 1,3-dipolarophiles, and Michael acceptors. They, as dienes, undergo a wide variety of synthetically useful Diels-Alder reactions and occasionally undergo even [4+6] cycloadditions. Many important reactions, which are not included in these categories, are also known. The present article describes the chemistry of thiophene 1,1-dioxides, with special emphasis being placed on their synthesis and synthetic applications. Many references on structural and theoretical studies are also given. The chemistry selenophene 1,1-dioxides is referred simply.

Keywords: Thiophene, selenophene, oxidation, cycloaddition, Michael addition

Topics in Current Chemistry, Vol. 205
© Springer-Verlag Berlin Heidelberg 1999

1
Introduction

Thiophene 1,1-dioxides are important compounds in the fields of both heterocyclic and heteroatom chemistry [1]. They are also important both from synthetic and mechanistic points of view. Oxidation of thiophenes initially affords thiophene 1-oxides [2], which are further oxidized to give the corresponding thiophene 1,1-dioxides as the final product. Although thiophenes are typical five-membered heteroaromatic compounds, thiophene 1,1-dioxides are no longer aromatic because all of the lone pair electrons on the sulfur atom are consumed for the bond formation with oxygen atoms. Thus, they now behave as unsaturated sulfones. Many of them are thermally labile because of easy dimerization and are highly reactive toward other substrates. The parent thiophene 1,1-dioxide and monosubstituted derivatives have eluded isolation until very recently. Monocyclic thiophene dioxides and benzo[b]thiophene dioxides serve as a 2π-component toward dienes and 1,3-dipoles, and monocyclic derivatives also act as dienes toward numerous 2π- and 6π-components. In addition, they undergo a wide variety of reactions with many other reagents. These are the very properties that make them synthetically most versatile.

The authors have attempted to make an exhaustive literature survey. The present review covers the whole field of chemistry of thiophene 1,1-dioxides, although special emphasis is placed on their synthetic use.

2
Synthesis

2.1
By Oxidation of Thiophenes

Oxidation of thiophenes provides the most convenient and simplest synthesis of the corresponding thiophene 1,1-dioxides (Scheme 1). Many oxidizing reagents

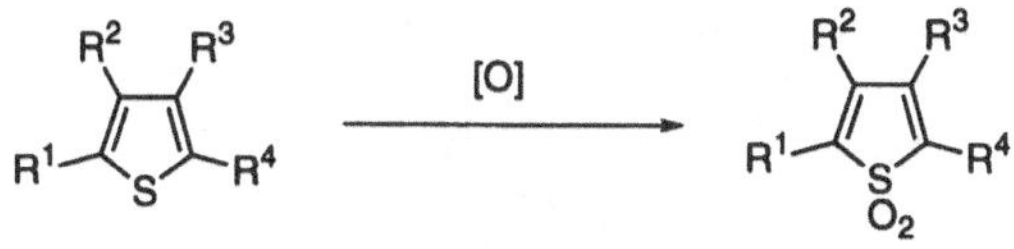

Scheme 1

have been used for this purpose. Organic peracids are the most common reagents. Peracetic, perbenzoic, m-chloroperbenzoic, and trifluoroperacetic acids, and, in rare cases, p-nitroperbenzoic and monoperphthalic acids, have been used. Recently, dimethyldioxirane has been shown to be very useful for this oxidation. Many inorganic oxidizing reagents are also available for this purpose. Among them, $NaBO_3 \cdot 4H_2O/AcOH$ and $F_2/H_2O/CH_3CN$ produce a variety of thiophene dioxides in better yields.

Perbenzoic acid is the oxidizing reagent that was used at the earliest stage of the chemistry of thiophene 1,1-dioxides. The thiophene 1,1-dioxides prepared by this oxidation are summarized in Table 1 [3–6]. Yields are also shown for convenience from a synthetic point of view throughout this review, if they were given in the literature.

Table 1. Preparation of thiophene 1,1-dioxides by oxidation with perbenzoic acid

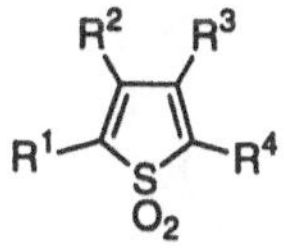

R^1	R^2	R^3	R^4	Yield (%)	Reference
H	Ph	Ph	H	40	3
H	4-MeC$_6$H$_4$	4-MeC$_6$H$_4$	H	34	4
Me	H	H	Me	14	5
H	Me	Me	H	62.5	5
Ph	H	H	Ph	31	5
COMe	Ph	Ph	H	72	5
COPh	Ph	Ph	H	48.5	5
COPh	Ph	Ph	COMe	60	5
Br	Ph	Ph	H	11	5
Br	Me	Me	Br		5
H	4-ClC$_6$H$_4$	4-ClC$_6$H$_4$	H	18	6

Peracetic acid is another oxidizing reagent used in the early stages. The oxidation has usually been carried out by use of a mixture of aqueous hydrogen peroxide and acetic acid [7–28], except in a rare case where a solution of peracetic acid in acetic acid was the reagent of choice [9]. The oxidation is easy to carry out and the reagents are inexpensive. Therefore, a great number of benzo-[9–17] and dibenzo-annelated [18–24] thiophene dioxides, which are much more stable than monocyclic derivatives, were prepared by this oxidation, although two papers reported the synthesis of monocyclic thiophene dioxides [7, 8]. Thiophene dioxides prepared in this way are summarized in Tables 2–4. Many thiophene dioxides with fused heterocyclic ring(s) were also synthesized by this method [25–28].

Table 2. Preparation of monocyclic thiophene 1,1-dioxides by oxidation with peracetic acid

$$R^2 \quad R^3$$
$$R^1 \underset{O_2}{\overset{S}{\diamond}} R^4$$

R^1	R^2	R^3	R^4	Reference
Ph$_2$CH	H	H	PhCH$_2$	7
Ph	H	H	Ph	8
4-CH$_3$C$_6$H$_4$	Cl	Cl	Ph	8
4-CH$_3$C$_6$H$_4$	Cl	Cl	4-CH$_3$C$_6$H$_4$	8
4-ClC$_6$H$_4$	Cl	Cl	Ph	8

Table 3. Preparation of benzo[b]thiophene 1,1-dioxides by oxidation with peracetic acid

$$R^3 \quad R^2$$
$$R^4 \qquad R^1$$
$$R^5 \underset{R^6 \; O_2}{\overset{S}{\diamond}}$$

R^1	R^2	R^3	R^4	R^5	Yield (%)	Reference
Me	Me	H	H	NO$_2$	79	9
H	Me	NO$_2$	H	H	85	10
Me	H	Br	H	Br		10
H	CH$_2$OEt	H	H	H	73	11
H	CH$_2$CN	H	H	H	74	11
H	CH$_2$Cl	H	H	H	16	11
2-(1-amino)naphthyl	Cl	H	H	H	71	11
2-(1-amino)naphthyl	Br	H	H	H	67	11
1-naphthyl	H	H	H	H		12
1-naphthyl	Br	H	H	H		12
1-naphthyl	NO$_2$	H	H	H		12
H	Me	H	H	H	70	13
H	H	H	Me	OMe		14
Ph	H	H	H	H		15
H	Ph	H	H	H		15
Ph	H	H	Me	H		15
H	H	H	Me	H		15
H	H	H	OMe	OMe		16
H	H	H	OMe	OMe		16
H	H		Br	NO$_2$	H	17

No substituent on the ring carbon unless otherwise stated.

Table 4. Preparation of dibenzothiophene 5,5-dioxides by oxidation with peracetic acid

R^1	R^2	R^3	R^4	Yield (%)	Reference
Br	H	H	Br	96	18
Et	H	H	Et	79	19
H	H	H	H	89	20
H	–CH=CH–CH=CH–		H		21
H	H	Me	H	66	22
H	NO$_2$	Me	H	17	22
H	NHAc	H	H	70	23
NO$_2$	H	I	OMe		24
NHAc	H	H	OMe		24

No substituent on the ring carbon unless otherwise stated.

Perfluoroacetic acid, which is highly electrophilic, can even oxidize electron-deficient thiophenes, such as perfluorodibenzothiophene and the related thiophenes, to the corresponding dioxides [29–31]. Benzo[b]thiophene 1,1-dioxide was also prepared in 58% yield by this oxidation [32].

m-Chloroperbenzoic acid (m-CPBA) is the reagent of recent, most frequent use [33–42]. With this reagent, a large number of thiophene 1,1-dioxides, such as tetrachloro- [35], 2,3- [38] and 3,4-di-tert-butyl- [37], 3,4-dineopentyl- [40], 3,4-

Table 5. Preparation of thiophene 1,1-dioxides by oxidation with m-CPBA

R^1	R^2	R^3	R^4	Yield (%)	Reference
Me	H	H	Me	52	33
t-Bu	H	H	t-Bu	70	33
t-Bu	H	t-Bu	H	56	33
Ph	H	H	Ph	74	33
Me	Me	Me	Me		34
Me	Ph	Ph	Me		34
Cl	Cl	Cl	Cl	50	35
C_nH_{2n+1}[a]	H	H	C_nH_{2n+1}[a]	35–51	36
H	t-Bu	t-Bu	H	93	37
t-Bu	t-Bu	H	H	79	38
H	1-ad[b]	1-ad[b]	H	75	39
H	neop[c]	neop[c]	H	68	40
t-Bu	t-Bu	t-Bu	t-Bu	75	41

[a] n = 6–11,   [b] 1-ad = 1-adamantyl,   [c] neop = neopentyl.

di(1-adamantyl)- [39], and tetra-*tert*-butylthiophene [41] 1,1-dioxides, were synthesized (Table 5). These dioxides are very important compounds both from the synthetic and structural points of view.

p-Nitroperbenzoic acid [43] and monoperphthalic acid [44] were used for preparation of dibenzothiophene dioxides.

Recently, dimethyldioxirane has been found to be very useful for oxidation of thiophenes (Table 6) [45]. The oxidation is carried out under neutral conditions. In addition, workup procedure is very simple since the dimethyldioxirane is converted into acetone. The oxidation is applicable to thiophenes carrying electron-withdrawing substituent(s), which resist oxidation with peracids. The reagent was also successfully applied to the oxidation of thiophenophanes [45]. The authors have succeeded in synthesis and isolation of the thermally unstable, parent thiophene 1,1-dioxide with this reagent [46].

Table 6. Preparation of thiophene 1,1-dioxides by oxidation with dimethyldioxirane

R^1	R^2	R^3	R^4	Yield (%)
Me	H	H	Me	93
$PhCH_2$	H	H	$PhCH_2$	93
Ph	Ph	Ph	Ph	99
Br	H	H	Br	27
COPh	Ph	Ph	COPh	76
Et	H	H	COMe	53

Table 7. Preparation of thiophene 1,1-dioxides by oxidation with inorganic oxidizing reagents

R^1	R^2	R^3	R^4	Yield (%)	Method[a]
Me	H	H	Me	78	A
C_9H_{19}	H	H	C_9H_{19}	65	A
– CH = CH – CH = CH–	H	H	95	A
– CH = CH – CH = CH–	– CH = CH – CH = CH–	95	A
Me	H	H	Me	95	B
Br	H	H	Br	95	B
Cl	H	H	Cl	70	B
Me	H	H	CO_2Et	90	B
– CH = CH – CH = CH	H	H	100	B

[a] Method A: $NaBO_3 \cdot H_2O$, AcOH, 45–50 °C. Method B: F_2, H_2O, CH_3CN, –10 °C.

Inorganic oxidizing reagents have also often been used. Sodium perborate tetrahydrate ($NaBO_3 \cdot 4H_2O$), which is commercially available, can oxidize a variety of thiophenes in acetic acid to the corresponding dioxides in good yields (Table 7) [47]. Treatment of thiophenes with molecular fluorine in a mixture of water and acetonitrile also affords the corresponding dioxides in excellent yields (Table 7) [48]. Other inorganic reagents used for the preparation of thiophene dioxides include $Na_2Cr_2O_7/H^+$ [49, 50], $KMnO_4$ [51], Ru on alumina, O_2 (70 atm) [52], $MoO_5/HMPA$ [53], and $NaOCl/H^+$ [54].

2.2
From Dihydrothiophene and Tetrahydrothiophene 1,1-Dioxides

The parent thiophene 1,1-dioxide 1 has been generated by dehydrobromination of the bromide 2 (Scheme 2) [55–57]. The reagents used for this conversion include piperidine [55], R_2CuLi, RLi, RMgBr, $BuNH_2$ [56], and K/ultrasound [57]. The use of K/ultrasound system in toluene allowed the formation of 3-methyl-, 3-chloro-, and 3,4-dimethylthiophene 1,1-dioxides [57]. Hofmann elimination of the ammonium salt 4 led to the parent thiophene dioxide [58]. Treatment of compound 3 with bases also generates the parent thiophene dioxide by elimination of two molecules of HBr [57, 59, 60].

Di- and tetrasubstituted thiophene dioxides 6 were prepared by dehydrohalogenation of the precursor compounds 5 (Scheme 3) [61–64]. For the prepara-

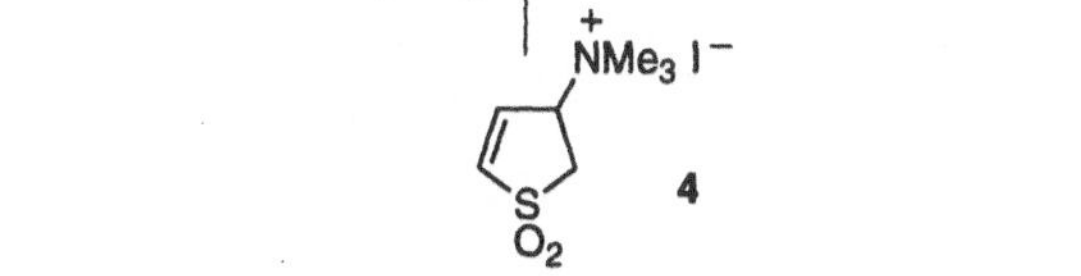

Scheme 2

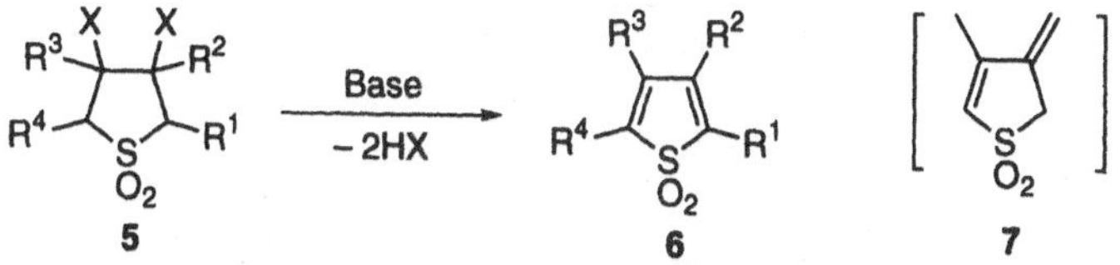

X = Cl or Br

a: $R^2 = R^4 = H$, $R^1 = R^3 = t$-butyl
b: $R^1 = R^4 = H$, $R^2 = R^3 = Cl$; 82%
c: $R^1 = R^4 = H$, $R^2 = R^3 = Me$
d: $R^1 = R^2 = R^3 = R^4 = Me$

Scheme 3

tion of 3,4-dimethylthiophene 1,1-dioxide, the isomeric exo-methylene compound **7** may be formed, depending upon the reaction conditions [62, 63].

Benzo[*b*]thiophene 1,1-dioxides could be also prepared by the following reductive methods (Scheme 4) [65–67].

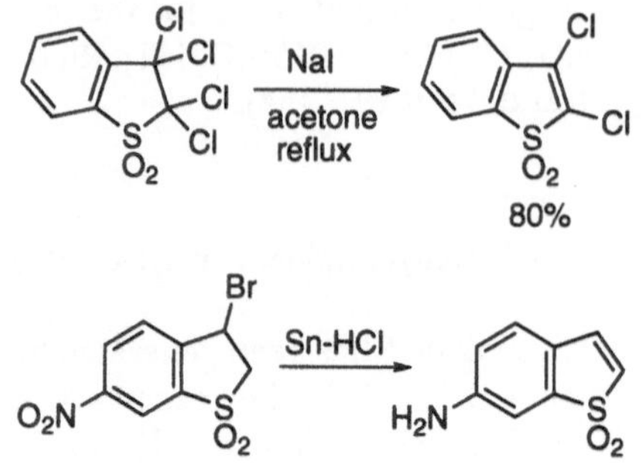

Scheme 4

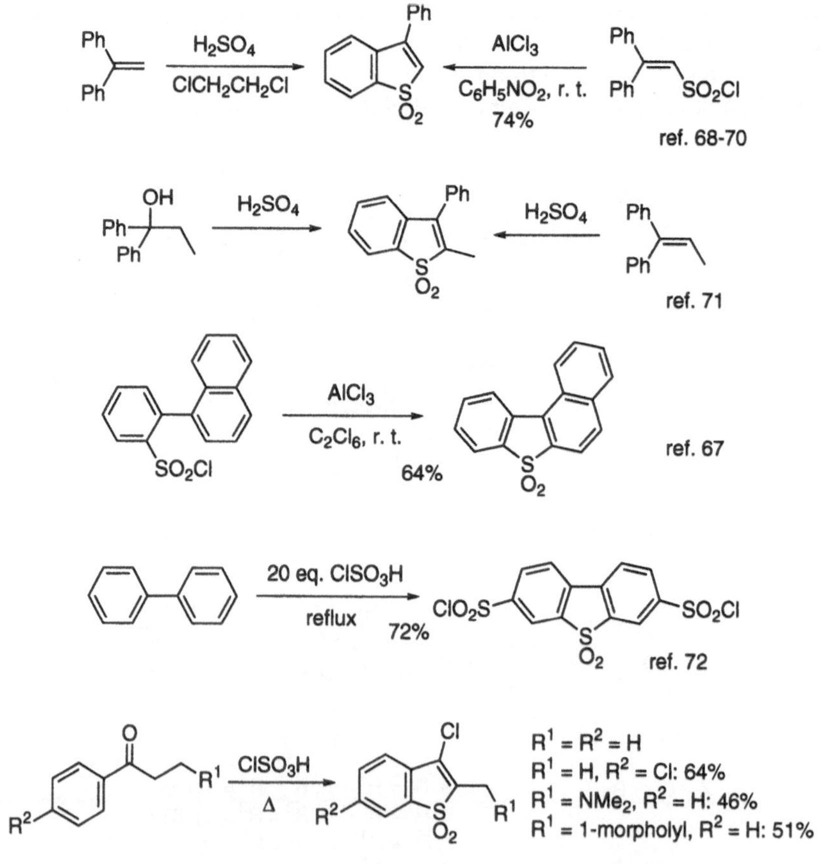

Scheme 5

ref. 73,74

2.3
By Intramolecular Cyclization

Intramolecular Friedel-Crafts type reactions of sulfonyl chlorides and related compounds often provide a facile route to benzo[*b*]thiophene 1,1-dioxides (Scheme 5) [67–74]. Several examples are summarized below.

An intramolecular cyclization probably involving a vinyl cation intermediate provides a convenient synthesis of benzo[*b*]thiophene 1,1-dioxides (Scheme 6) [75, 76].

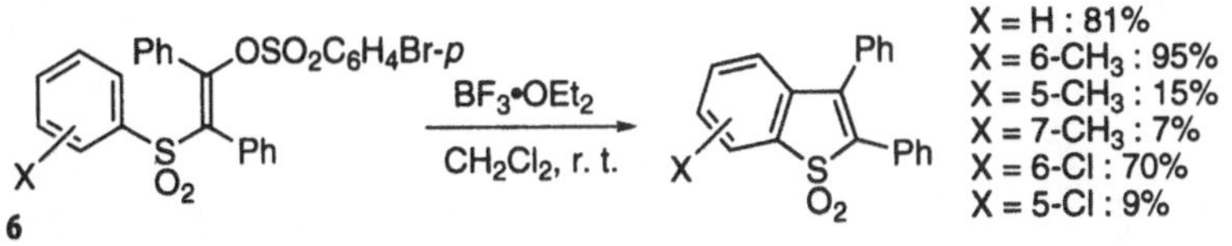

X = H : 81%
X = 6-CH$_3$: 95%
X = 5-CH$_3$: 15%
X = 7-CH$_3$: 7%
X = 6-Cl : 70%
X = 5-Cl : 9%

Scheme 6

Photostimulated reactions of *o*-bis(phenylsulfonyl)benzene derivatives with sodium arenethiolates in DMSO afford dibenzothiophene 5,5-dioxides in moderate yields along with other products (Scheme 7) [77, 78]. Electrochemical reduction of the above substrates also leads to dibenzothiophene 5,5-dioxides [79–81]. Both reactions involve (*o*-phenylsulfonyl)phenyl radical intermediates.

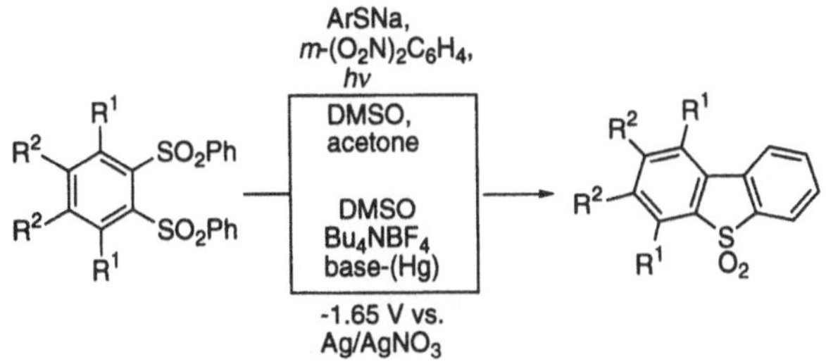

Scheme 7

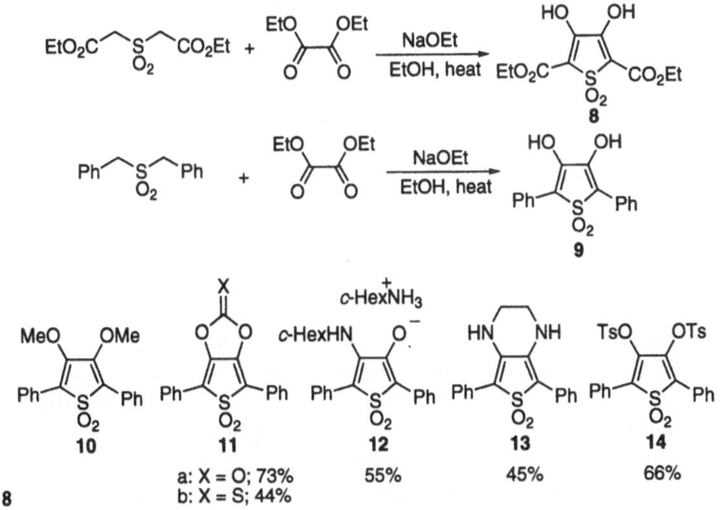

Scheme 8

Hinsberg type condensation, though the initial step is not intramolecular, affords 3,4-dihydroxythiophene 1,1-dioxides such as **8** and **9** (Scheme 8) [82]. The latter compound was converted into a series of derivatives **10–14** [83–85].

Other examples of condensation reactions and related reactions [86–89] leading to thiophene 1,1-dioxides are summarized below (Scheme 9).

Scheme 9   R = H, Me; Ar = C$_6$H$_5$, 4-ClC$_6$H$_4$, 3-FC$_6$H$_4$, 4-FC$_6$H$_4$, 4-MeOC$_6$H$_4$

2.4
Miscellaneous

The following examples involve thiophene 1,1-dioxide formation by miscellaneous methods (Scheme 10) [90–96]. See also [97–100], although they appear synthetically less important.

3
Reactivities

3.1
Cycloadditions and Related Reactions

3.1.1
[1+2] Cycloaddition

Although the behavior of thiophenes toward carbenes has been fairly well documented, the reaction of thiophene 1,1-dioxides with carbenes or carbenoids has scarcely been reported. The dibromide **3** reacted under the conditions given below to produce 3-dihalomethylenedihydrothiophene dioxide **15** in reasonable yields (Scheme 11) [101]. The reaction may involve the dihalocarbene adducts **16** of the parent thiophene 1,1-dioxide, but the reaction more probably proceeds

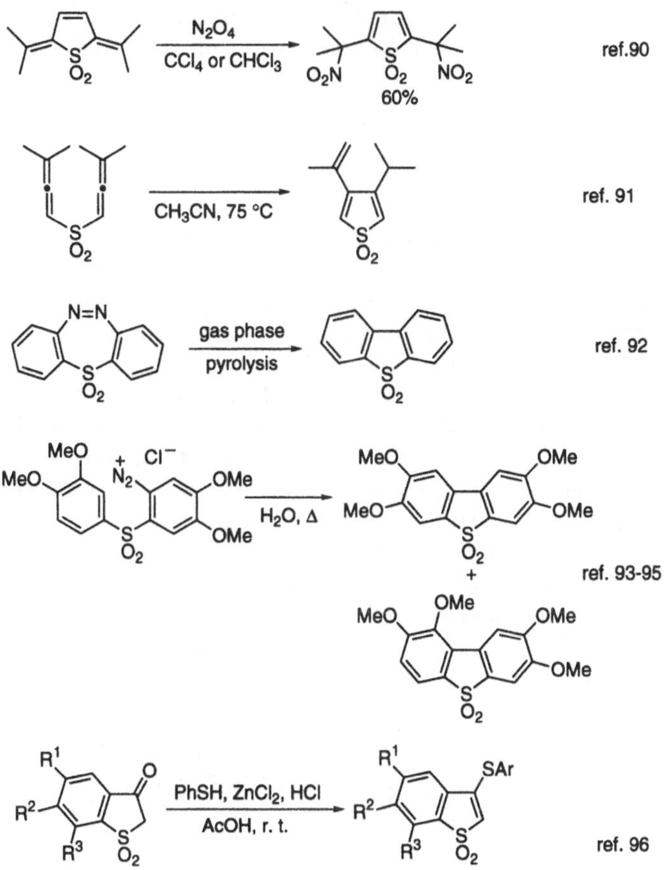

ref.90

ref. 91

ref. 92

ref. 93-95

ref. 96

$R^1 = R^2 = R^3 = H$, Ar = Ph : 81%
$R^1 = Me$, $R^2 = R^3 = H$, Ar = Ph : 79%
$R^1 = Me$, $R^2 = R^3 = H$, Ar = p-Tol : 72%
$R^1 = R^2 = H$, $R^3 = Me$, Ar = Ph : 74%
$R^1 = R^2 = H$, $R^3 = Me$, Ar = p-Tol : 71%
$R^1 = R^2 = H$, $R^3 = Cl$, Ar = Ph : 82%
$R^1 = R^2 = H$, $R^3 = Cl$, Ar = p-Tol : 76%

Scheme 10

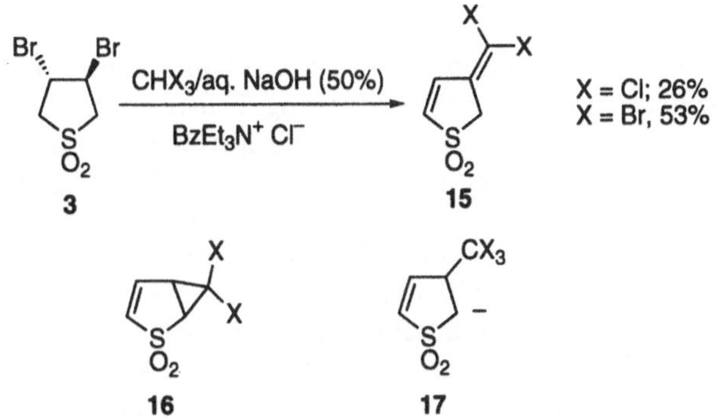

X = Cl; 26%
X = Br, 53%

Scheme 11

through Michael addition of trihalomethyl anions to the dioxide. The Michael addition of nucleophiles to thiophene 1,1-dioxides is described in full detail later.

3.1.2
[2+2] and [2+4] Cyclodimerization

Thiophene 1,1-dioxides undergo cyclodimerization in two ways. Many monocyclic thiophene 1,1-dioxides are thermally labile and undergo [2+4] dimerization, one molecule acting as a dienophile and the other acting as a diene. On the other hand, thermally stable thiophene 1,1-dioxides, particularly benzo[*b*] thiophene 1,1-dioxides, undergo [2+2] dimerization on photoirradiation.

Thus, the parent thiophene 1,1-dioxide **1** undergoes [2+4] dimerization to form compound **18** with spontaneous loss of SO_2 from the initial adduct [55, 102]. Compound **18** may undergo further cycloaddition with **1** and an appropriate dienophile such as maleic anhydride to give **19a** and/or **19b** and **20**, respectively (Scheme 12). This is the very property that had made isolation of **1** difficult. Very recently, the authors have succeeded in isolation and full characterization of **1**, and its half-life in a dilute solution (0.13 mol dm^{-3}) was determined to be ca. 140 min at 25 °C [46].

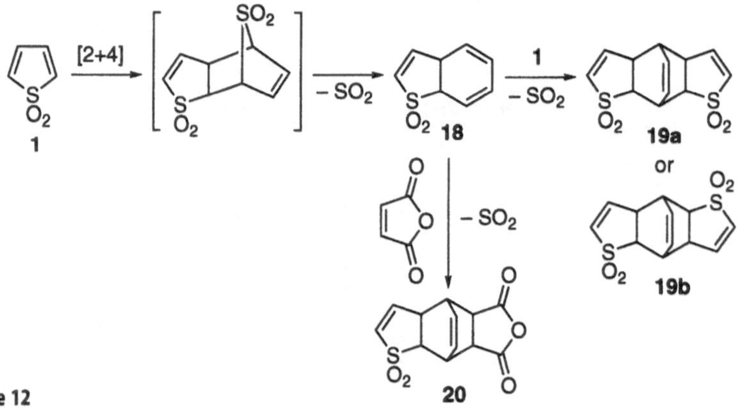

Scheme 12

Thermal stability of thiophene 1,1-dioxides increases with an increasing number of substituents on the thiophene ring. Thus, dimerization of 3,4-dichlorothiophene 1,1-dioxide **21** requires heating in refluxing xylene to give compound **22** in 49% yield with elimination of SO_2 and HCl along with compounds **23** ([2+2] head-to-tail adduct) and **24** ([2+4] adduct of **21** and **22**) (Scheme 13) [103]. Dimerization of the more thermally stable dioxide **25** requires more forcing conditions (reflux in phenol) to give **26** in 62% yield (Scheme 14) [104]. Interestingly, heating **27** in refluxing *t*-BuOH for a prolonged period of time affords pentasubstituted benzenes **29** in good yields with spontaneous loss of SO_2 and HX (X=Cl, Br) from the initial adduct **28** (Scheme 15) [105].

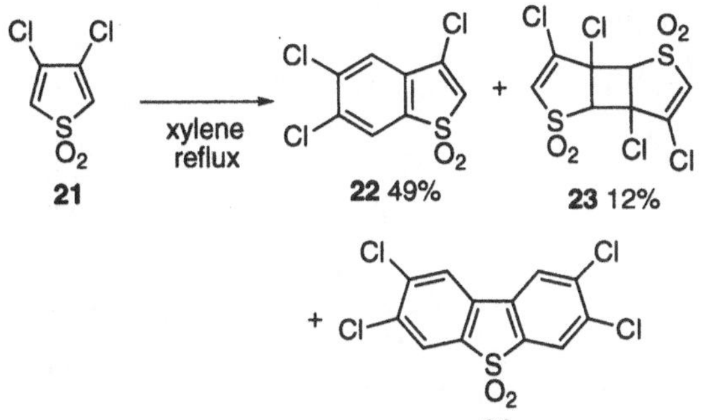

Scheme 13

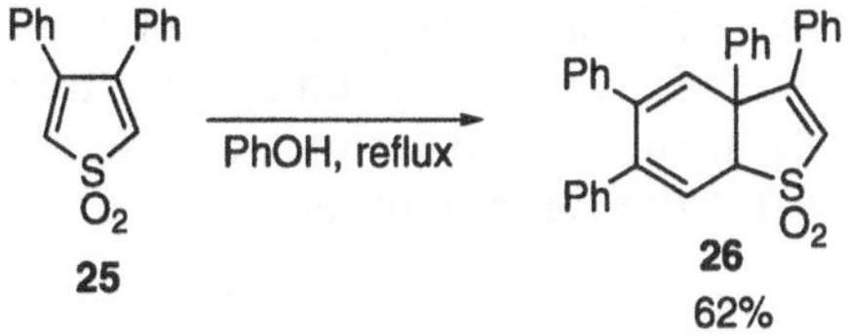

Scheme 14

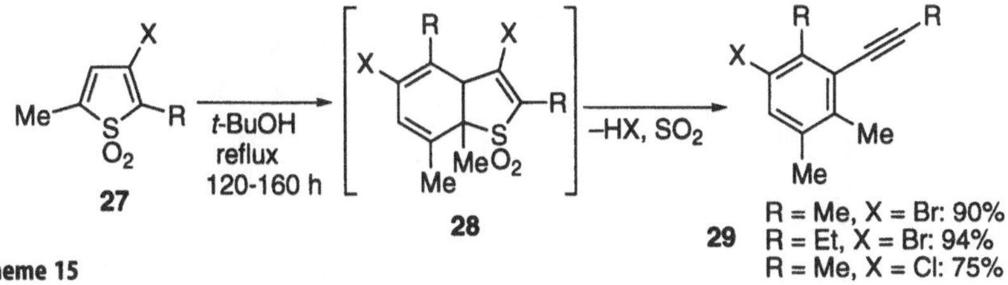

Scheme 15

When 4-bromo-2-methyl-5-trimethylsilylthiophene was oxidized with *m*-CPBA and the resulting dioxide **30** was purified on a column of Al_2O_3, desilylation took place to give **31**, which dimerized to give isomeric products **32** and **33** in 46% and 22% yields, respectively (Scheme 16) [106]. Similarly, the dioxide **34** was also desilylated on a column of Al_2O_3, and the resulting dioxide **35** spontaneously dimerized to give **36** in 56% yield. Treatment of **36** with a base such as EtMgBr afforded **37** in 89% yield (Scheme 17) [107]. Similar transformations starting from some other 2-trimethylsilyl substituted thiophenes were reported [107].

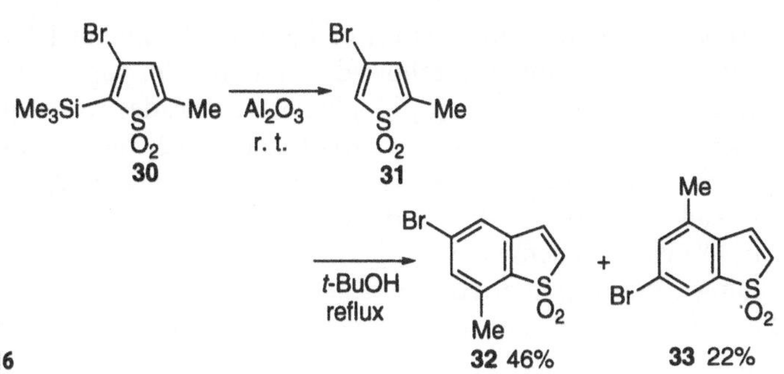

Scheme 16

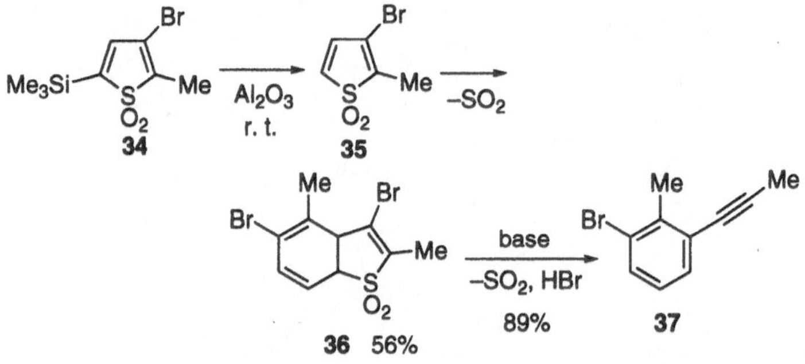

Scheme 17

Even more stable benzo[b]thiophene 1,1-dioxides undergo [2+4] dimerization under forcing conditions. Thus, heating the parent dioxide **38** under forcing conditions affords compounds **39** or **40**, depending upon the reaction conditions (Scheme 18) [67, 108–111]. Such dimerization was also reported with several derivatives of **38** [67, 112]. A typical example is the formation of **42** and **43** from **41** (Scheme 19) [67].

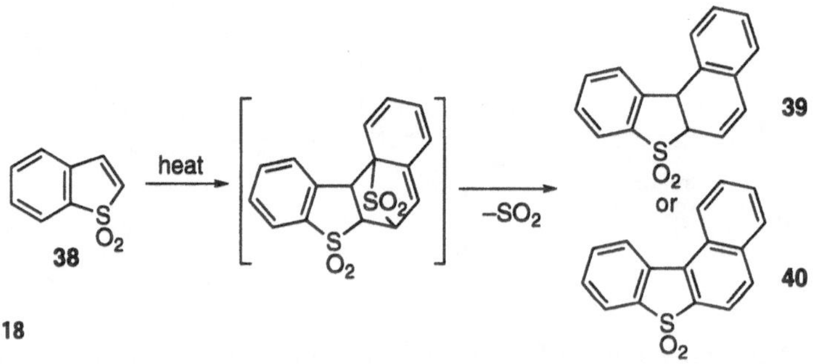

Scheme 18

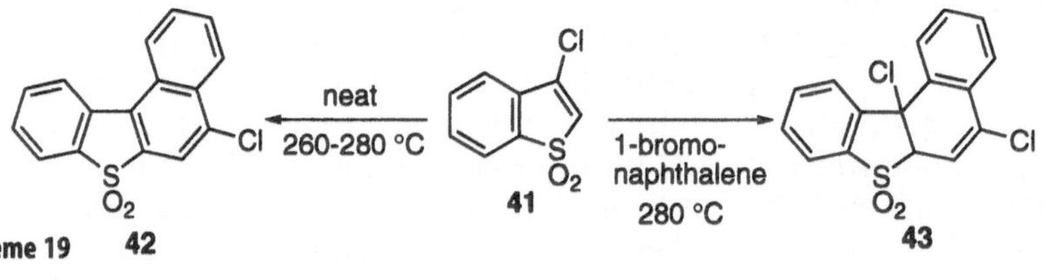

Scheme 19   **42**

The first report on photodimerization of benzo[b]thiophene 1,1-dioxides appeared in 1955 [113, 114], and was followed by a number of papers [115–119]. When a benzene solution of benzo[b]thiophene 1,1-dioxides was exposed to sunlight for several days, good yields of mixtures of [2+2] head-to-head and head-to-tail dimers were formed [113–115]. In the case of the parent dioxide **38**, *anti* head-to-head and *anti* head-to-tail dimers, **44** and **45**, were obtained in the ratio 73:27 in 7% yield (Scheme 20) [116].

The foregoing dimerization often results in the exclusive formation of either a head-to-head or a head-to-tail isomer. Several examples are given in Scheme 21 [117–119].

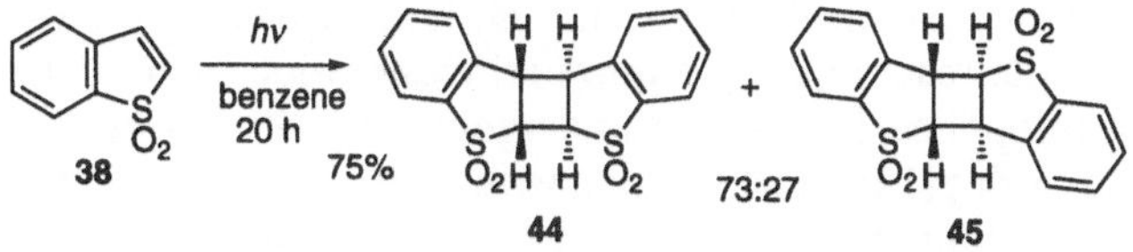

Scheme 20

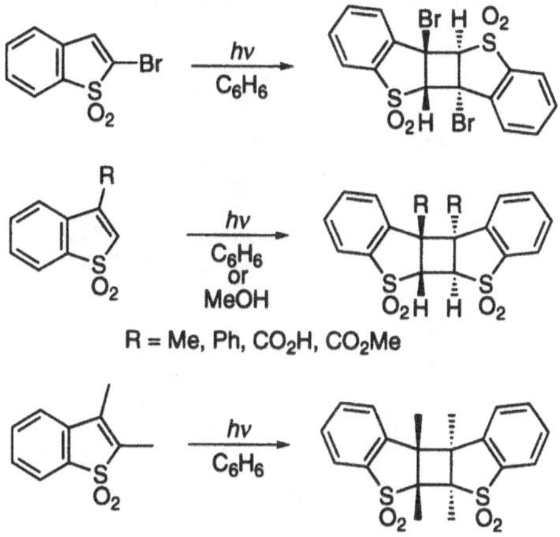

R = Me, Ph, CO₂H, CO₂Me

Scheme 21

3.1.3
[2+2] Photocycloaddition

Benzo[*b*]thiophene 1,1-dioxide **38** undergoes [2+2] photocycloaddition with vinyl acetate [120] and 1-(diethylamino)propyne [121] to give the corresponding adducts in excellent yields (Scheme 22). The reaction appears synthetically important, and its scope and limitation deserve examination in detail.

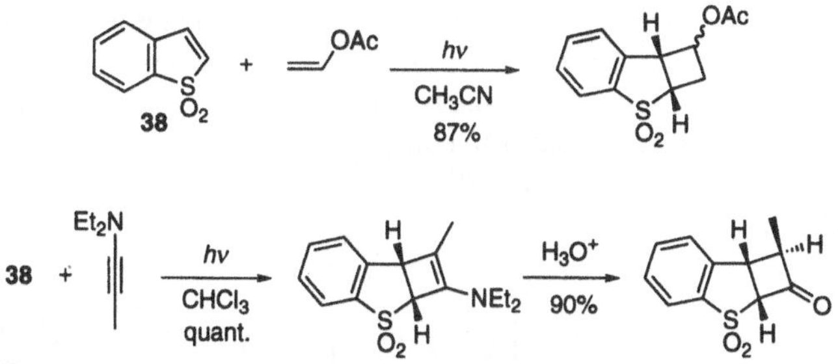

Scheme 22

3.1.4
1,3-Dipolar Cycloaddition

Thiophene 1,1-dioxides undergo 1,3-dipolar cycloadditions with a variety of 1,3-dipoles [122–131]. The cycloaddition is regioselective in most cases. Thus, it was claimed that 3,4-diphenyl- and 3,4-di-*p*-tolylthiophene 1,1-dioxides react with diazomethane to give the adducts **46**, which eliminate nitrogen to give compounds **47** as the final product (Scheme 23) [122]. However, the reported regiochemistry of the reaction may require reinvestigation by modern techniques. The parent thiophene 1,1-dioxide undergoes a 1,3-dipolar cycloaddition with benzonitrile oxides [123]. The reaction with an equimolar amount of nitrile oxides affords the mono-adducts in a regioselective manner, and the reaction with two molar amounts of nitrile oxides produces the bis-adducts (Scheme 24).

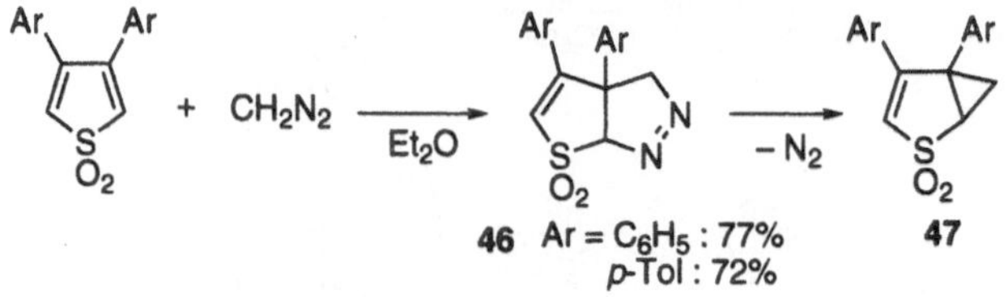
+ CH2N2
Et2O
− N2
46 Ar = C6H5 : 77%
p-Tol : 72%
47

Scheme 23

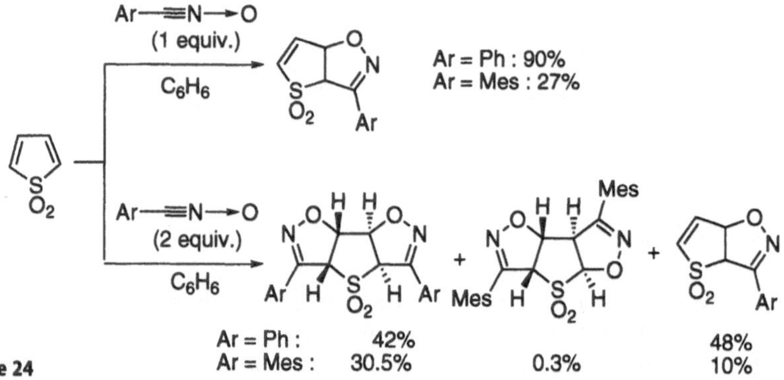
Ar—≡N—→O
(1 equiv.)
C6H6
Ar = Ph : 90%
Ar = Mes : 27%
Ar—≡N—→O
(2 equiv.)
C6H6
Ar = Ph : 42% 48%
Ar = Mes : 30.5% 0.3% 10%

Scheme 24

A report in 1974 claimed that benzo[*b*]thiophene 1,1-dioxide **38** reacted with diazomethane at room temperature to give the adduct **48** in a good yield, which isomerized to **49** on heating [124]. However, a more recent report proposed the isomeric structure **50** to this adduct based on ^{1}H NMR analyses (Scheme 25) [125]. The adduct **50** was converted into compound **51** on photolysis [126]. Also reported was the reaction of the dioxide **52** with methyldiazomethane (Scheme 26) [125].

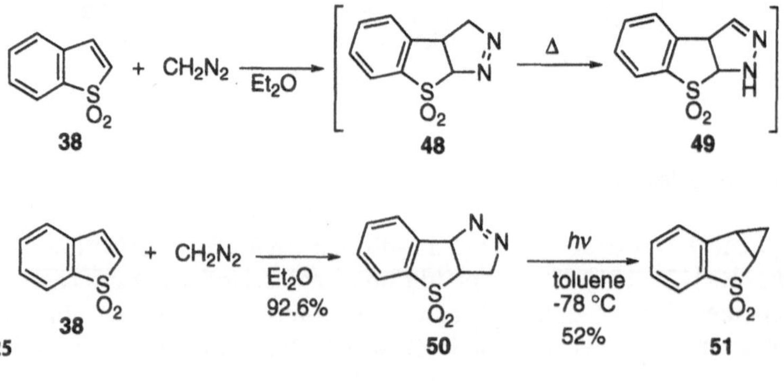
+ CH2N2
Et2O
Δ
38
48
49
+ CH2N2
Et2O
92.6%
hv
toluene
-78 °C
52%
38
50
51

Scheme 25

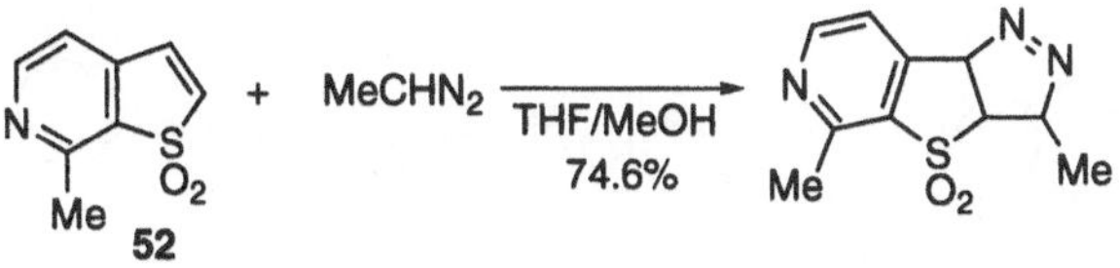
+ MeCHN2
THF/MeOH
74.6%
Me
Me
Me

Scheme 26 52

Benzo[*b*]thiophene 1,1-dioxide **38** and its many derivatives undergo 1,3-dipolar cycloadditions with a variety of 1,3-dipoles such as nitrile oxides (Scheme 27) [123, 127, 128], nitrones (Scheme 28) [127], and nitrile imines (Scheme 29) [129, 130].

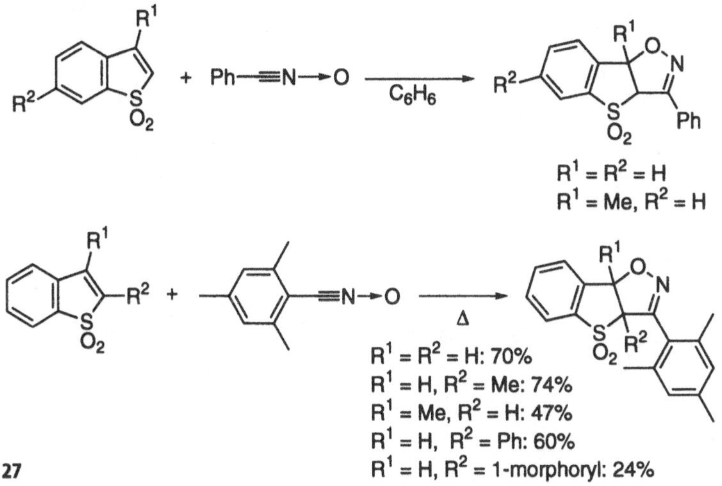
+ Ph—≡N→O
C6H6
R1 = R2 = H
R1 = Me, R2 = H
Δ
R1 = R2 = H: 70%
R1 = H, R2 = Me: 74%
R1 = Me, R2 = H: 47%
R1 = H, R2 = Ph: 60%
R1 = H, R2 = 1-morphoryl: 24%

Scheme 27

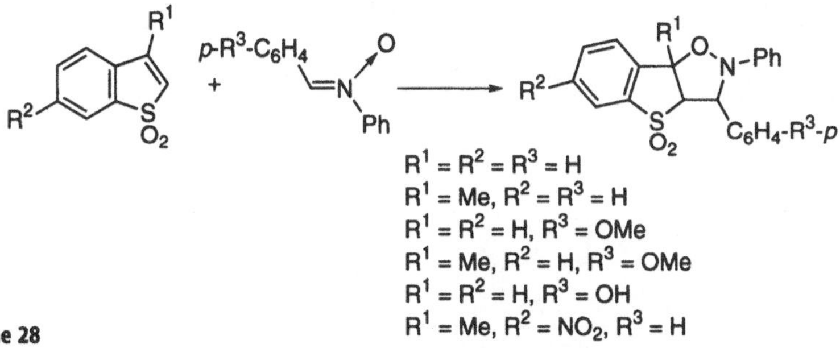
p-R3-C6H4
Ph
N-Ph
C6H4-R3-p
R1 = R2 = R3 = H
R1 = Me, R2 = R3 = H
R1 = R2 = H, R3 = OMe
R1 = Me, R2 = H, R3 = OMe
R1 = R2 = H, R3 = OH
R1 = Me, R2 = NO2, R3 = H

Scheme 28

3.1.5
[2+4] Cycloaddition (Thiophene Dioxides as 2π Component)

Monocyclic thiophene 1,1-dioxides generally act as a 4π component toward dienophiles. However, in some cases, thiophene 1,1-dioxides such as parent thiophene, 3,4-dichlorothiophene, and tetrachlorothiophene dioxides, **1** [132], **21** [133], and **53** [134], respectively, behave as a dienophile toward 4π components (Scheme 30). Examples are shown below.

As a typical cyclic α,β-unsaturated sulfone, benzo[*b*]thiophene 1,1-dioxide **38** participates in [2+4] cycloadditions (Scheme 31) [135].

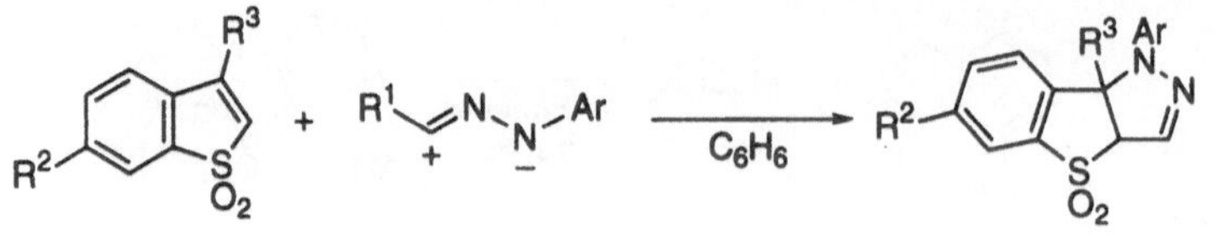

Scheme 29

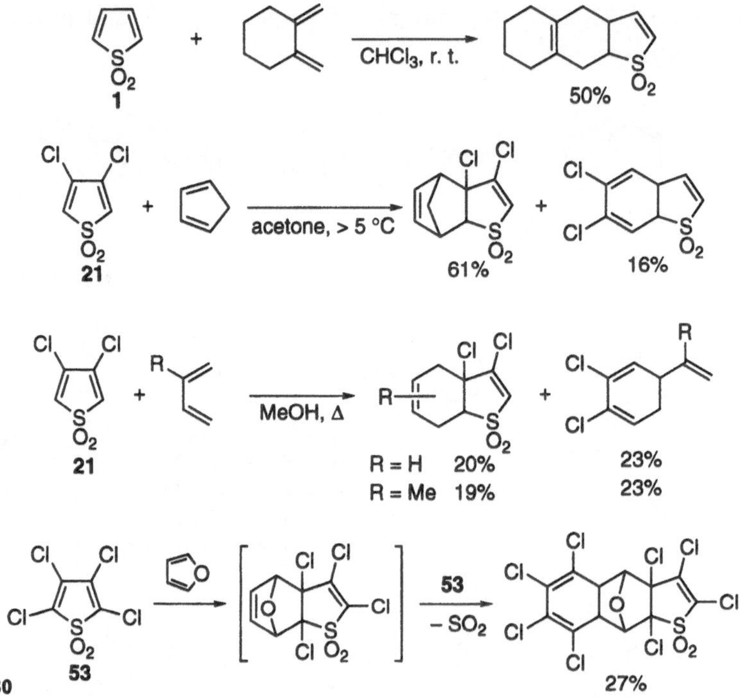

Scheme 30

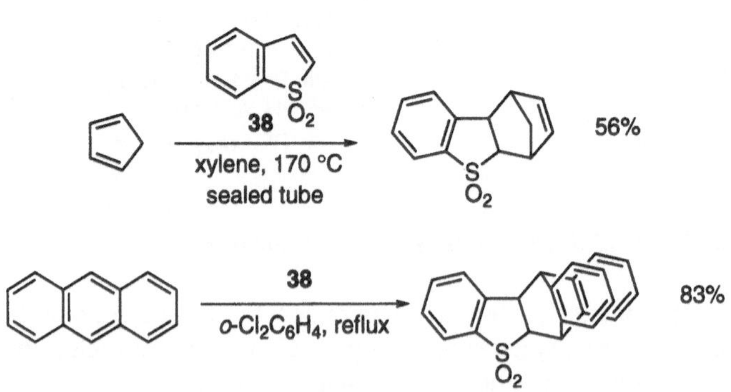

Scheme 31

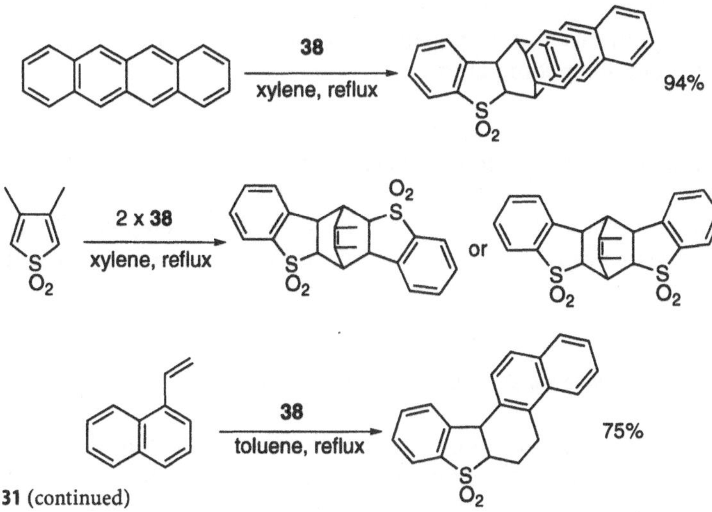

Scheme 31 (continued)

Derivatives of **38** such as **22** [133], **54** [135], and **55** [136] (Fig. 1) also act as 2π components toward dienes.

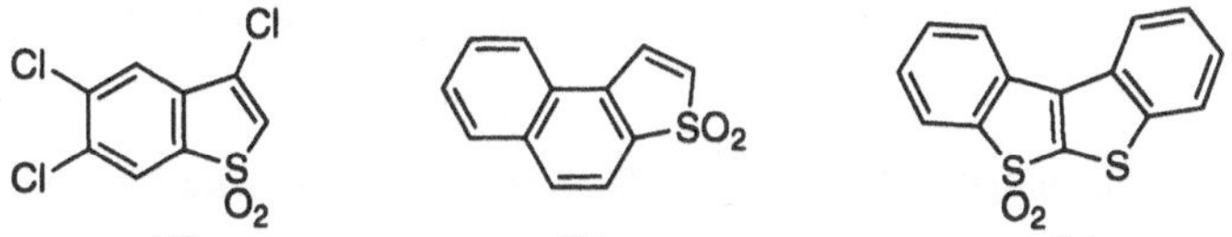

Fig. 1. Some derivatives of **38** which also act as 2p components toward dienes

3.1.6
[4+2] Cycloaddition (Thiophene Dioxide as 4π Component)

Thiophene dioxides act as a 4π component toward a wide variety of dienophiles. Because of the electron-withdrawing properties of the sulfonyl group, they are rather electron-deficient dienes. Therefore, dienophiles do not necessarily require activation by electron-withdrawing group(s) in order to undergo cyclo-addition with most thiophene dioxides. The electron demand of the Diels-Alder reactions of thiophene dioxides often becomes inverse to that of common Diels-Alder reactions. Even electron-rich alkenes can then take part in the Diels-Alder reactions with thiophene dioxides. This is one of the features that makes thio-phene dioxides synthetically important. Although the Diels-Alder reactions of many thiophene dioxides have been examined in great detail, the most extensi-vely investigated are those with tetrachlorothiophene 1,1-dioxide **53**. This is easily obtainable by oxidation of tetrachlorothiophene and is thermally stable, but is very reactive and can react with a wide variety of alkenic dienophiles.

Recently, thiophene dioxides carrying bulky substituents at 3- and 4-positions have been synthesized and applied to the preparation of congested benzene derivatives and other compounds.

3.1.6.1
[4+2] Cycloaddition with Alkenic Dienophiles

In the present review, [4+2] cycloadditions with alkenic dienophiles are classified for convenience into six categories according to the structures of the final products.

3.1.6.1.1
Category A: 1,3-Cyclohexadiene-Forming Reactions

Diels-Alder reactions of thiophene 1,1-dioxides usually lead to 1,3-cyclohexadiene derivatives with spontaneous loss of sulfur dioxide from the initial adducts. This type of reaction has been most extensively studied with tetrachlorothiophene 1,1-dioxide **53**. Several reactions with **53** are summarized in Table 8 [134].

The excellent report on the dioxide **53** by Raasch [134] was followed by a number of similar papers [139–156]. Table 9 summarizes these results.

3,4-Dichlorothiophene 1,1-dioxide **21** reacts with a series of 1,4-dihydronicotinates **56** in refluxing toluene to give the corresponding tetrahydroquinolines

Table 8. Preparation of tetrachlorocyclohexadiene derivatives with tetrachlorothiophene 1,1-dioxide (**53**) [135]

Dienophiles	Products	Yields
$H_2C=CH_2$		89%
		$n = 0: 74\%$ $n = 1: 90\%$ $n = 2: 92\%$ $n = 3: 79\%$ $n = 7: 40\%$
		$m = n = 1: 88\%$ $m = 4, n = 2: 63\%$

Table 8 (continued)

Dienophiles	Products	Yields
		$R, R = CH_2CH_2CH_2$:75 % $R, R = CH_2(CH_2)_3CH_2$:73 % $R, R = CH_2CH_2C(=CH_2)CH_2CH_2$:75 %
		74 %
		27 %
		80 %
		91 %
		65 %
		$R = CH_2CH_2C°CH$:70 % $R = CH_2(CH_2)_8CH_3$:88 % $R = CH_2(CH_2)_3C=CH_2$:69 % $R = CH_2(CH_2)_4C=CH_2$:69 % $R = CO_2H$:68 % $R = CO_2Me$:91 % $R = CN$:83 % $R = CH_2CO_2H$:78 % $R = CH_2(CH_2)_7CO_2H$:78 % $R = CH_2Br$:88 % $R = CH_2NCS$:55 % $R = 4\text{-}(1,2\text{-methylenedioxy})benzyl$:75 % $R = 2\text{-pyridyl}$:75 % $R = N\text{-pyrrolidonyl}$:84 %

Table 8 (continued)

Dienophiles	Products	Yields
		65%
		X = NH : 88% X = NMe : 90% X = O : 61%
		74%
		68%
		X = O : 92% X = H₂ : 84%
		77%
		58%
		74%

Table 9. Preparation of tetrachlorocyclohexadiene derivatives with tetrachlorothiophene 1,1-dioxide (53)

Dienophiles	Products	Yields
		69% [137]
		99% [138]
		$R^1 = H, R^2 = Cl$: 37% $R^1 = Cl, R^2 = H$: 40% $R^1 = R^2 = Cl$: 45% [139, 140]
		91% [141]
		91% [142, 143]
		$X = NCO_2Et$: 66% $X = CH_2$: 85% $X = O$: 92% [144]
		quant. [145, 146]
		R, R = H, H: 97% R, R = OCH_2CH_2O: 86% [147]

Table 9 (continued)

Dienophiles	Products	Yields
		R, R = H, H: 94% R, R = OCH$_2$CH$_2$O: 87% [147]
		58–77% [148]
		92% [148]
		[148]
		[149]
		58% [150]
		[150]

Table 9 (continued)

Dienophiles	Products	Yields
		[150]
		54% [151]
		96% [152]
		[153]
		[154]
		n = 0: 85% n = 1: 96% n = 2: 85% [155]
		n = 0: 95% n = 1: 96% [155]
		34% [156]

57 regio- and stereoselectively in reasonable yields; in many cases, the reaction was carried out in the presence of 2,6-lutidine to capture hydrogen chloride evolved from self-dimerization product of 21 (Scheme 32) [157]. The reactions that belong to the present category were investigated practically only with 53 and 21 [158, 159].

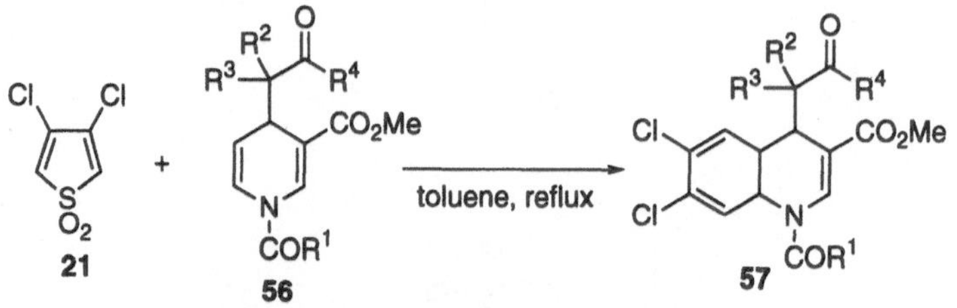

R^1 = OEt, R^2 = R^3 = H, R^4 = Ph : 21%
R^1 = OEt, R^2 = H, R^3 = Ph, R^4 = OMe : 52%
R^1 = OEt, R^2 = R^3 = Me, R^4 = OMe : 36%
R^1 = NMe$_2$, R^2 = H, R^3 = Ph, R^4 = OMe : 54%
R^1 = NMe$_2$, R^2 = R^3 = Me, R^4 = OMe : 36%
R^1 = OEt, R^2 = R^3 = H, R^4 = Ph : 61%[a]
R^1 = OEt, R^2 = R^3 = H, R^4 = p-MeOC$_6$H$_4$: 21%[a]
R^1 = OMe, R^2 = H, R^3 = Me, R^4 = Et : 55%[a]
R^1 = OMe, R^2 = H, R^3 = Me, R^4 = Ph: 31%[a]
R^1 = OMe, R^2 = R^3 = R^4 = Me : 22%[a]
R^1 = NMe$_2$, R^2 = R^3 = R^4 = Me : 22%[a]

Scheme 32   [a] Results in the presence of 2,6-lutidine

3.1.6.1.2
Category B: Benzene Ring-Forming Reactions by Removal of a Small Molecule from Initial Adducts

Cycloaddition of thiophene dioxides with alkenic dienophiles often directly leads to the formation of a benzene ring. This can be attained by extrusion of sulfur dioxide from the initial adducts, followed by 1) elimination of a small molecule, 2) an intramolecular hydrogen transfer, and 3) (accidental) oxidative dehydrogenation under applied conditions.

In the following examples, sulfur dioxide and benzenesulfenic acid (benzene-sufinic acid) are simultaneously extruded from the initial adducts. In this way, a series of highly congested benzene derivatives, which are otherwise difficult to prepare, were synthesized in good overall yields (Scheme 33) [39, 40, 159, 160].

In the following instances, dibromobenzocyclobutadiene, generated in situ, reacted with thiophene dioxides to give the initial adducts 58, which in turn were debrominated to give the biphenylene derivatives 59 in modest yields (Scheme 34) [161].

3,4-Dichlorothiophene 1,1-dioxide 21 reacted with p-benzoquinone to give the naphthoquinone 61 directly in an excellent yield because the primary adduct 60 was dehydrogenated by p-benzoquinone (Scheme 35) [133]. For the reaction

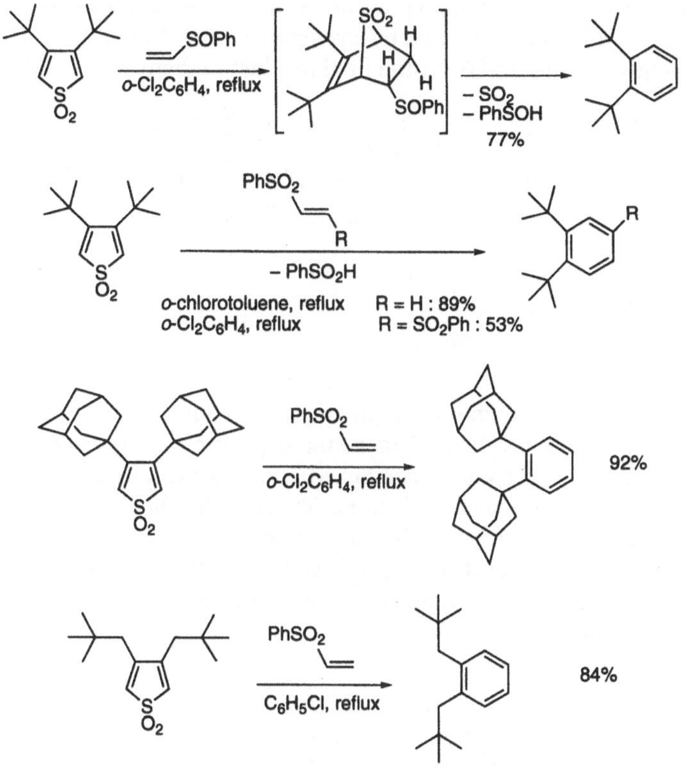

Scheme 33

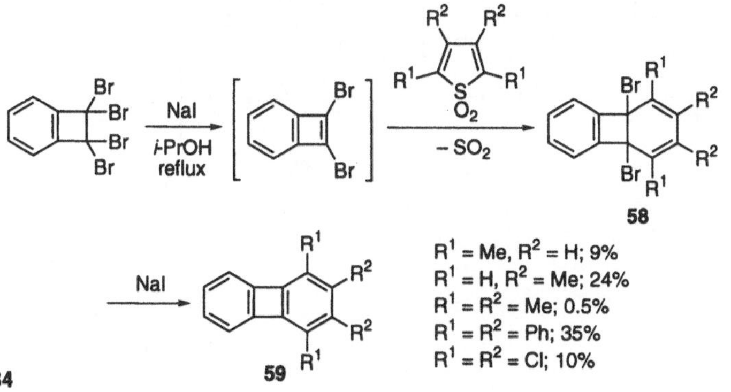

R[1] = Me, R[2] = H; 9%
R[1] = H, R[2] = Me; 24%
R[1] = R[2] = Me; 0.5%
R[1] = R[2] = Ph; 35%
R[1] = R[2] = Cl; 10%

Scheme 34

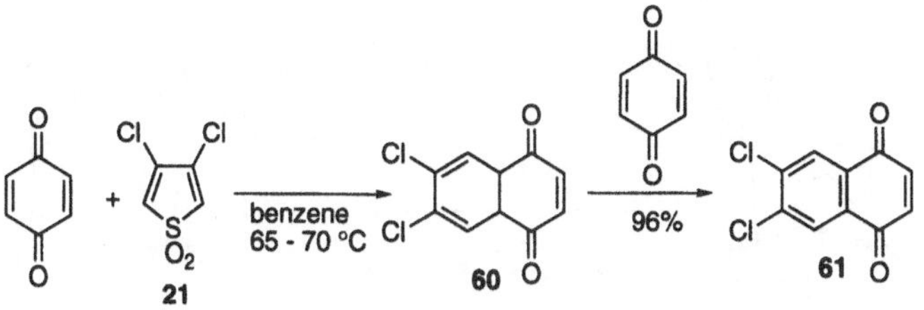

Scheme 35

with norbornadiene, the primary adduct **62** undergoes a retro Diels-Alder reaction to give *o*-dichlorobenzene and cyclopentadiene. The latter further reacts with **21** to give compound **63** as the final product (Scheme 36) [133].

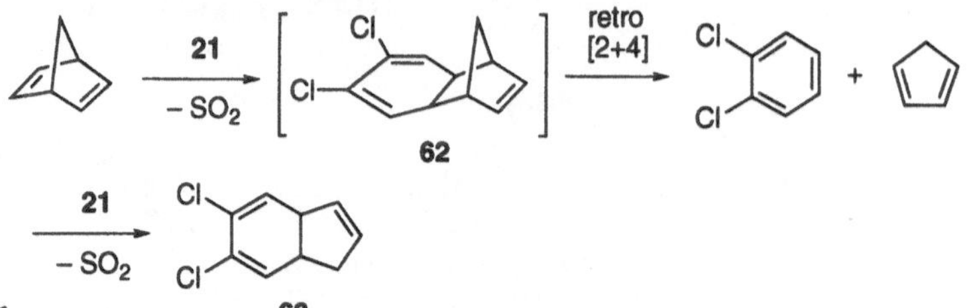

Scheme 36

The reaction of tetrachlorothiophene 1,1-dioxide **53** with a dienophile **64** affords the primary product **65** in a stereospecific manner because of steric demand [162, 163]. In **65**, the cyclohexadiene ring protons are placed in close proximity to the other ring double bond. This enables the hydrogen atom transfer to take place stereospecifically to give **66** (Scheme 37). In the case of the dienophile **67**, such a reaction is impossible, and the normal product **68** was obtained (Scheme 38). An additional two examples are also shown below (Scheme 39) [162–165].

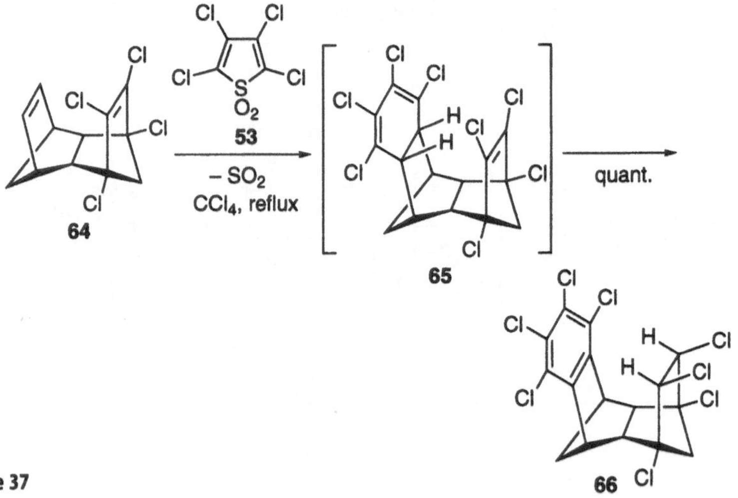

Scheme 37

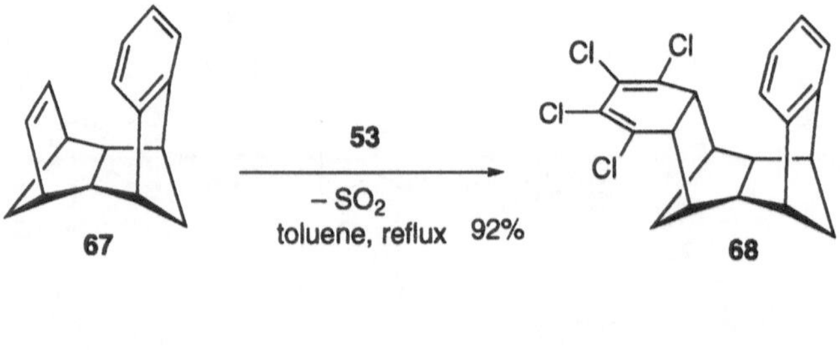

Scheme 38

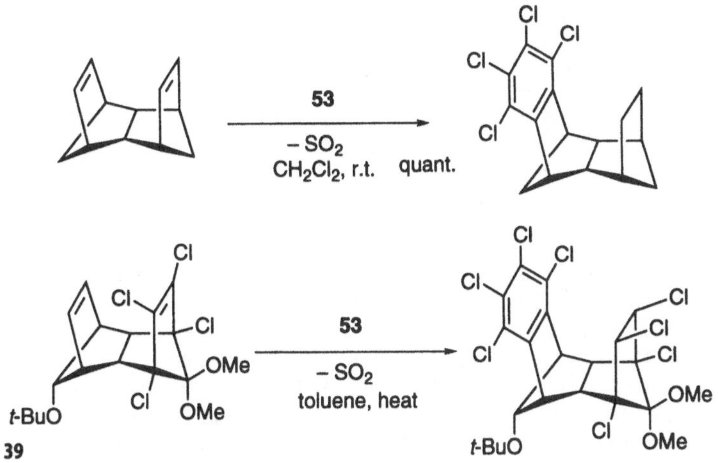

Scheme 39

In the following examples, the primary cyclohexadiene adducts are converted into the corresponding benzene derivatives by fortuitous dehydrogenation (Scheme 40) [132, 166–168].

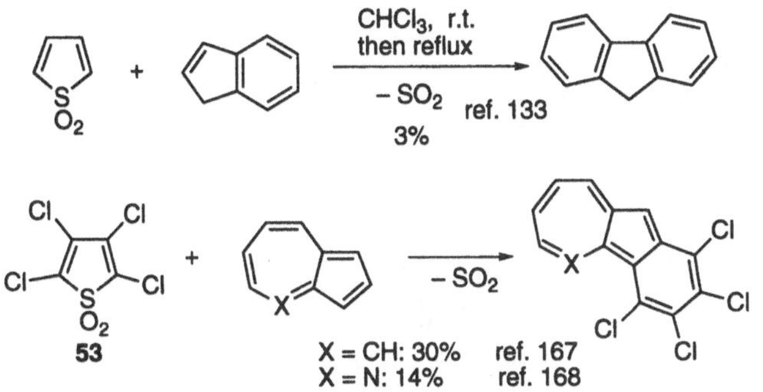

Scheme 40

3.1.6.1.3
Category C: Benzene Ring-Forming Reactions by Ring Opening of Initial Adducts

2,5-Dichloro-, tetrabromo-, and tetrachlorothiophene 1,1-dioxides react with a series of furans in regioselective manner to give the primary adducts **69**, which aromatize by ring opening to give benzene derivatives **70** in excellent yields (Scheme 41) [134]. The reaction of 3,4-di-*t*-butylthiophene 1,1-dioxide with 2,5-dimethylfuran also gives a highly congested benzene **71** (Scheme 42) [159]. The reaction of tetrachlorothiophene 1,1-dioxide **53** with the oxazole **72** gives the urea derivative **73** in a similar manner (Scheme 43) [168]. On the other hand, for the adduct with the oxazole **74**, such aromatization cannot occur, and so the exo methylene compound **75** is formed in an excellent yield by migration of a hydrogen of the methyl group (Scheme 44).

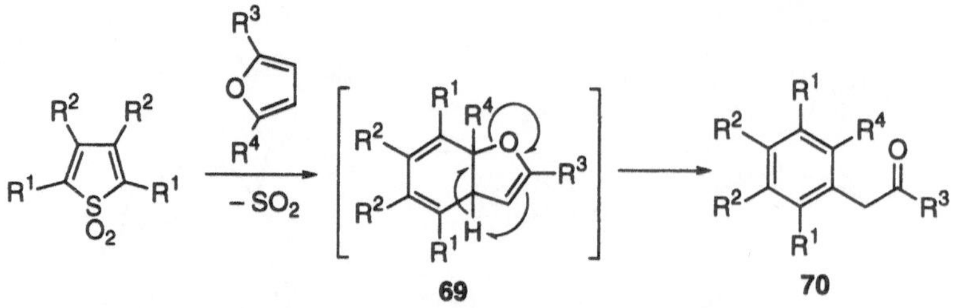

69   **70**

hexane, 65 °C, 30 min $R^1 = R^2 = Cl$, $R^3 = Me$, $R^4 = H$: 92%
hexane, 65 °C, 30 min $R^1 = R^2 = Cl$, $R^3 = Bu$, $R^4 = H$: 93%
93 °C, 30 min $R^1 = R^2 = Cl$, $R^3 = R^4 = Me$: 90%
93 °C, 30 min $R^1 = R^2 = Br$, $R^3 = R^4 = Me$: 90%
93 °C, 30 min $R^1 = Cl$, $R^2 = H$, $R^3 = R^4 = Me$: 77%
ClCH$_2$CH$_2$Cl, 83 °C, 1 h $R^1 = R^2 = Cl$, $R^3 = CH_2OAc$, $R^4 = H$: 74%
ClCH$_2$CH$_2$Cl, 83 °C, 1 h $R^1 = R^2 = Cl$, $R^3 = CH_2NHAc$, $R^4 = H$: 74%
CH$_2$Cl$_2$, 30 °C $R^1 = R^2 = Cl$, $R^3 = OMe$, $R^4 = H$: 77%
100 °C, 22 h $R^1 = R^2 = Cl$, $R^3 = CO_2Me$, $R^4 = H$: 62% (enol form)
125 °C, 1 h $R^1 = R^2 = Cl$, $R^3 = COMe$, $R^4 = H$: 37% (enol form)

Scheme 41

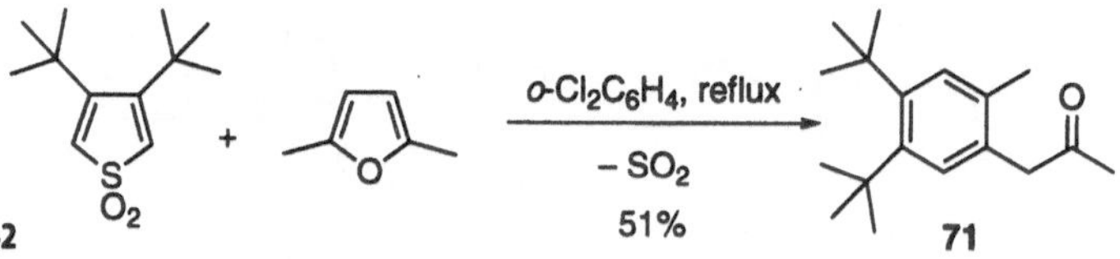

Scheme 42   51%   **71**

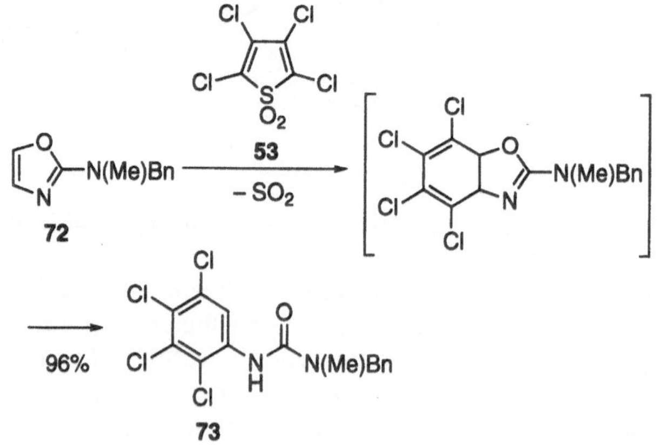

53

72

96%

Scheme 43   **73**

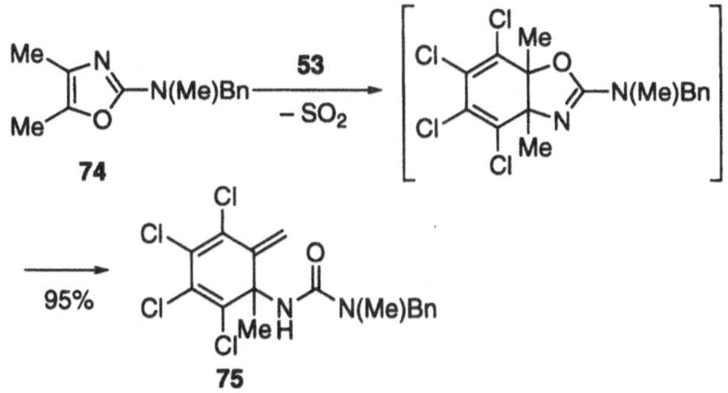

Scheme 44

3.1.6.1.4
Category D: Seven- and Eight-Membered Ring-Forming Reactions

A series of thiophene dioxides **6** react with a variety of cyclopropenes to give cycloheptatrienes **77** in reasonable yields (Scheme 45) [169–172]. In this case, spontaneous ring opening of the adducts is involved, though loss of sulfur dioxide and the ring opening of the primary adducts **76** may take place simultaneously. Benzocyclobutadiene, generated in situ, reacts with thiophene dioxides **6** to produce adducts **78**, which immediately undergo ring opening to give benzocyclooctatetraenes **79** in reasonable yields (Scheme 46) [173].

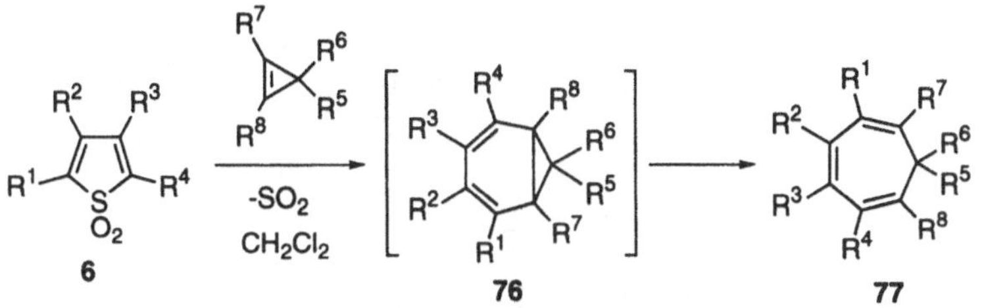

$R^1 = R^4 = Me, R^2 = R^3 = H$ $R^5 = R^6 = R^7 = R^8 = H$
$R^5 = Me, R^6 = R^7 = R^8 = H$
$R^5 = R^6 = R^8 = H, R^7 = Me$

$R^1 = R^4 = t\text{-Bu}, R^2 = R^3 = H$ $R^5 = Me, R^6 = R^7 = R^8 = H$
$R^1 = R^3 = t\text{-Bu}, R^2 = R^4 = H$ $R^5 = R^6 = R^7 = R^8 = H$
$R^5 = R^6 = R^8 = H, R^7 = Me$

$R^1 = R^3 = H, R^2 = R^4 = Me$ $R^5 = R^6 = R^8 = H, R^7 = Me$
$R^5 = Me, R^6 = R^7 = R^8 = H$

$R^1 = R^2 = R^3 = Me, R^4 = H$ $R^5 = Me, R^6 = R^7 = R^8 = H$
$R^1 = R^2 = R^3 = R^4 = Me$ $R^5 = R^6 = R^7 = R^8 = H$
$R^5 = R^6 = R^8 = H, R^7 = Me$
$R^5 = Me, R^6 = R^7 = R^8 = H$

Scheme 45 $R^1 = R^2 = R^3 = R^4 = H$ $R^5 = R^6 = H, R^7 = Cl, R^8 = Br$

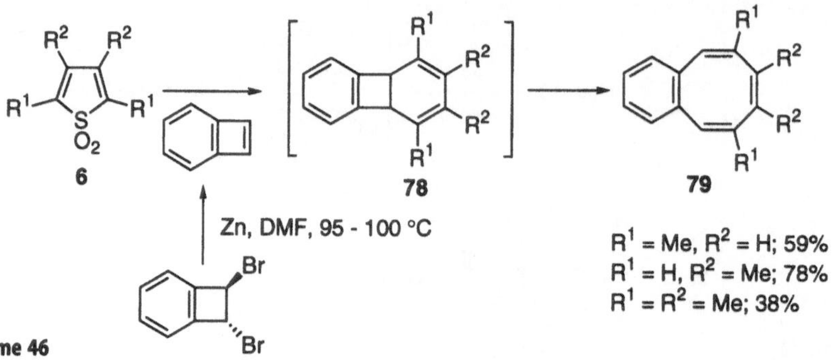
6
78
79
Zn, DMF, 95 - 100 °C
R¹ = Me, R² = H; 59%
R¹ = H, R² = Me; 78%
R¹ = R² = Me; 38%

Scheme 46

3.1.6.1.5
Category E: [4+2] Cycloaddition Followed by Intramolecular [4+2] Cycloaddition

Raasch showed that 3,4-dichloro-, tetrabromo-, and tetrachlorothiophene 1,1-dioxides react with a range of cyclic and acyclic non-conjugated dienes to give polycyclic compounds in good yields by [4+2] cycloaddition followed by intramolecular [4+2] cycloaddition [35]. For example, the Diels-Alder reaction of these dioxides with 1,5-cyclooctadiene produces the adducts **80**, which then undergo an intramolecular Diels-Alder reaction to give polycyclic compounds **81** in good yields in one pot (Scheme 47). In this way, a variety of polycyclic compounds, shown below, were synthesized (Table 10).

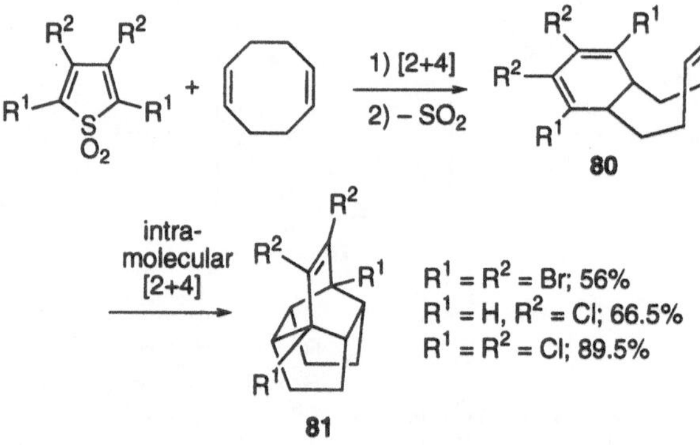
1) [2+4]
2) – SO2
80
intra-
molecular
[2+4]
81
R¹ = R² = Br; 56%
R¹ = H, R² = Cl; 66.5%
R¹ = R² = Cl; 89.5%

Scheme 47

Table 10. Preparation of polycyclic compounds with thiophene 1,1-dioxides

Starting dienes	Polycyclic products	Yields
		R = R' = H: 50% R = H, R' = Me: 48% R = R' = Me: 57%
		72%
		X = O, Y = CH$_2$, R = H: 80% X = S, Y = CH$_2$, R = H: 75% X = O, Y = CO, R = H: 56% X = O, Y = CO, R = H: 35%
		X = CH$_2$: 26% X = O: 65% X = S: 66% X = NAc: 59% X = NCN: 56% X = CHCN: 75%
		57%
		58%
		R^1 = R^2 = Br: 68% (tetrabromothiophene 1,1-dioxide as starting material) R^1 = H, R^2 = Cl: 17% (3,4-dichlorothiophene 1,1-dioxide as starting material)

The above reaction was applied to the preparation of structurally interesting cage compounds (82) (Scheme 48) [145, 146].

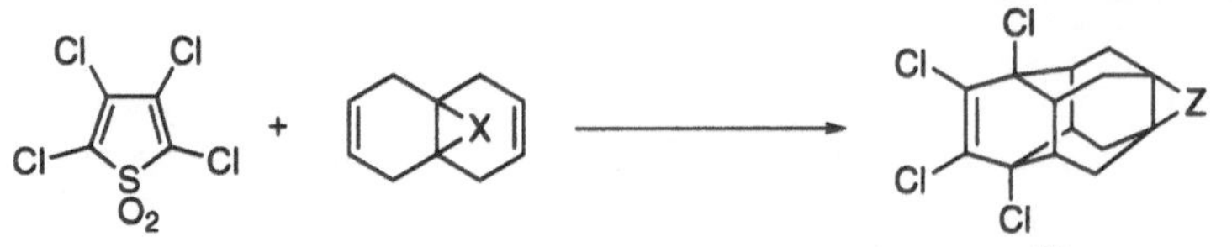

X = O	2,6-di-*t*-butyl-4-methylphenol, NaHCO$_3$, CCl$_4$, reflux	
X = CCl$_2$	neat, 120 °C	33%
X = NHCO	1,2-dichloroethane, reflux	66%
X = CO$_2$CO	toluene, reflux	70%
X = CH$_2$OCH$_2$	toluene, reflux	40%
X = OC(CH$_3$)O	K$_2$CO$_3$, toluene, reflux	41%

Scheme 48

3.1.6.1.6
Category F: Bis-Adduct Formation

The primary adducts, cyclohexadiene derivatives, formed by [4+2] cycloaddition of thiophene dioxides with dienophiles, may further undergo [4+2] cycloaddition with the dienophiles. Thus, the adducts 84 of 3,4-di-*tert*-butylthiophene dioxide 83 with maleic anhydride and *N*-phenylmaleimide further react with these dienophiles to give excellent yields of bis-adducts, which are composed of the endo-endo and endo-exo isomers, 85a and 85b (Scheme 49) [160]. A similar reaction was also observed with 3,4-dichlorothiophene 1,1-dioxide with *N*-butyl- and *N*-*p*-nitrophenylmaleimides (Scheme 50) [133]. The reaction of highly congested thiophene dioxides 87 with 4-phenyl-1,2,4-triazoline-3,5-dione provides a unique pyridazine synthesis since the bis-adducts 88 are converted into the corresponding pyridazines 89 in one pot and in good yields by treatment with KOH in methanol (Scheme 51) [174].

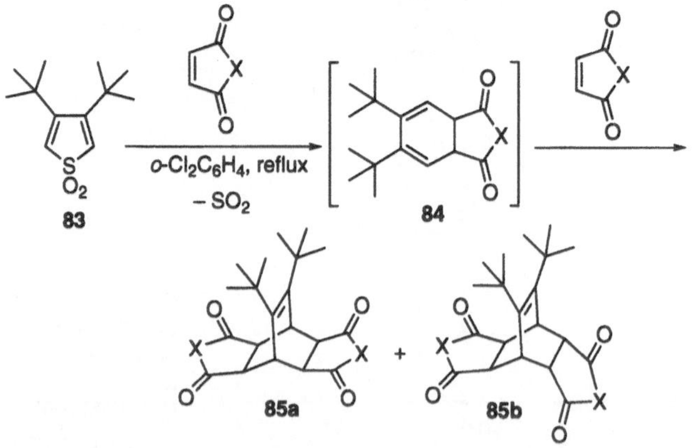

X = O : 73%   25%
X = NPh : 26%   40%

Scheme 49

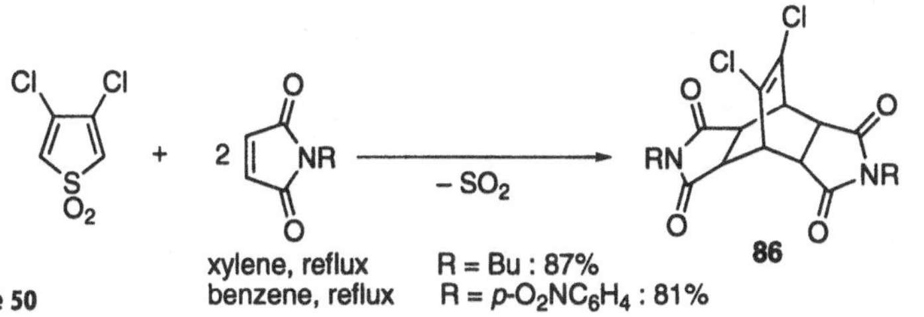

+ 2
NR
- SO2
RN
NR
86
xylene, reflux
benzene, reflux
R = Bu : 87%
R = p-O2NC6H4 : 81%

Scheme 50

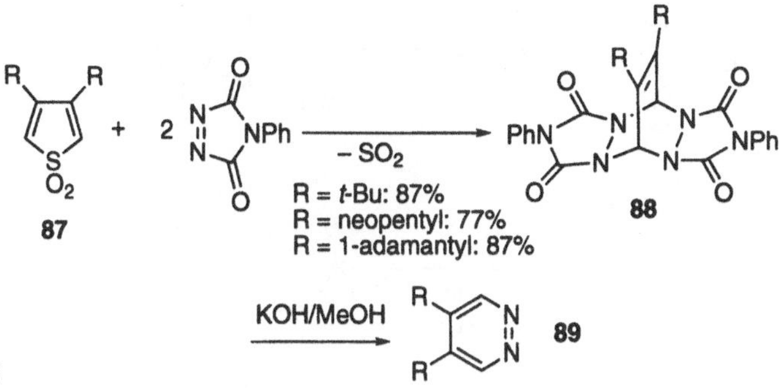

87
+ 2
NPh
- SO2
PhN
NPh
88
R = t-Bu: 87%
R = neopentyl: 77%
R = 1-adamantyl: 87%
KOH/MeOH
89

Scheme 51

In the following examples, the primary adducts **90** further reacted with the dioxide **53** to give the bis-adducts **91** (Scheme 52) [35].

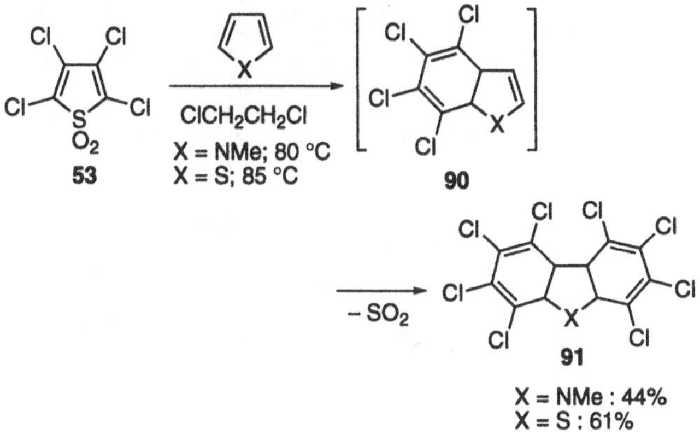

53
ClCH2CH2Cl
X = NMe; 80 °C
X = S; 85 °C
90
- SO2
91
X = NMe : 44%
X = S : 61%

Scheme 52

3.1.6.2
[4+2] Cycloaddition with Alkynic Dienophiles

[4+2] Cycloaddition of thiophene 1,1-dioxides with alkynic dienophiles leads to the formation of benzene derivatives with elimination of sulfur dioxide. Thus, the unstable parent thiophene 1,1-dioxide **1** reacts with diethyl acetylenedicarboxylate and cyclooctyne to give diethyl phthalate and benzocyclooctene, although in low yields (Scheme 53) [132, 175]. Cycloadditions with alkenic and alkynic dienophiles had been used as evidence for the generation of **1** until spectroscopic evidence became available [46]. Tetrachlorothiophene dioxide **53** reacts with phenylacetylene [35] and a cyclic alkyne **92** [176] to give 1,2,3,4-tetrachloro-5-phenylbenzene and compound **93**, respectively (Scheme 54).

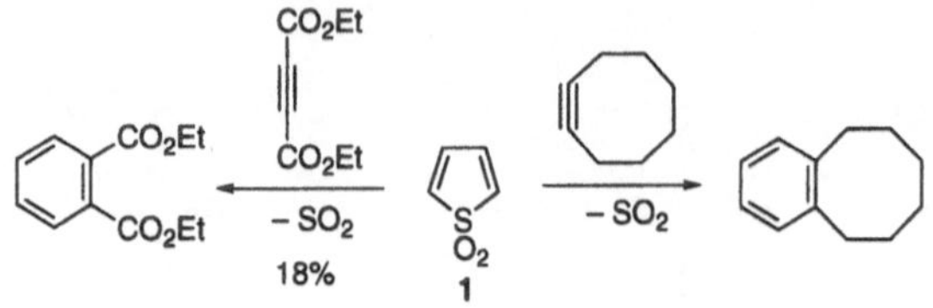
CO2Et
CO2Et
CO2Et
CO2Et
- SO2
18%
S
O2
1
- SO2

Scheme 53

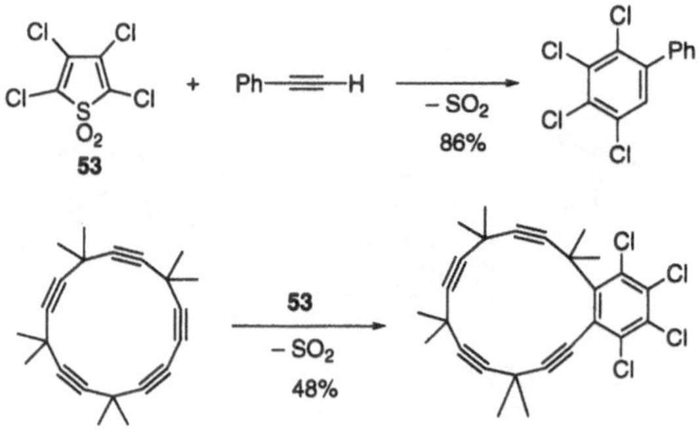
Cl
Cl
Cl
Cl
S
O2
53
+
Ph—≡—H
- SO2
86%
Cl
Cl
Ph
Cl
Cl
53
- SO2
48%
Cl
Cl
Cl
Cl

Scheme 54 92 93

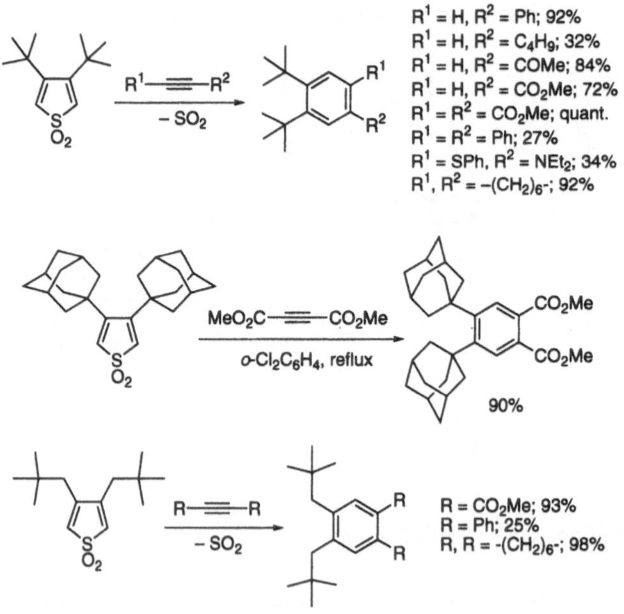
S
O2
R1—≡—R2
- SO2
R1
R2
R1 = H, R2 = Ph; 92%
R1 = H, R2 = C4H9; 32%
R1 = H, R2 = COMe; 84%
R1 = H, R2 = CO2Me; 72%
R1 = R2 = CO2Me; quant.
R1 = R2 = Ph; 27%
R1 = SPh, R2 = NEt2; 34%
R1, R2 = -(CH2)6-; 92%
S
O2
MeO2C—≡—CO2Me
o-Cl2C6H4, reflux
CO2Me
CO2Me
90%
S
O2
R—≡—R
- SO2
R
R
R = CO2Me; 93%
R = Ph; 25%
R, R = -(CH2)6-; 98%

Scheme 55

The synthetically most important reaction that belongs to this category would be [4+2] cycloadditions of highly congested thiophene dioxides with alkynic dienophiles (Scheme 55) [39, 40, 159]. These reactions afford *o*-di-*tert*-butyl-, *o*-dineopentyl-, and *o*-di(1-adamantyl)benzene derivatives in good yields, synthesis of which is very difficult by other means.

Benzyne and substituted benzynes, generated by thermal decomposition of 2-carboxybenzenediazonium chloride or by aprotic diazotization of appropriate anthranic acids, reacted with a variety of thiophene dioxides to produce naphthalene derivatives generally in moderate yields (Scheme 56) [34, 159]. In two cases, the naphthalenes so produced further reacted with benzyne to give benzobarrelene derivatives.

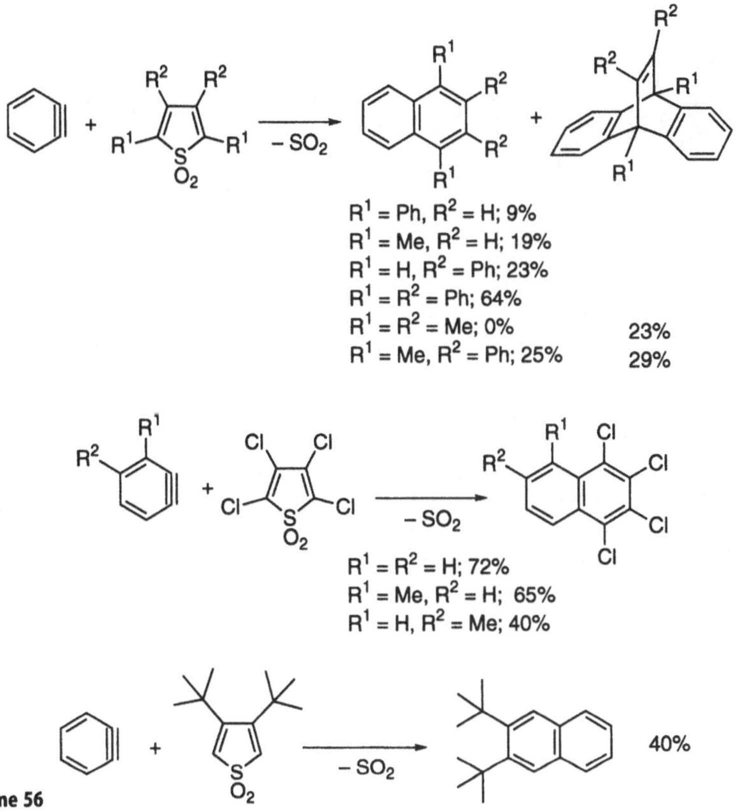

$R^1 = Ph, R^2 = H; 9\%$
$R^1 = Me, R^2 = H; 19\%$
$R^1 = H, R^2 = Ph; 23\%$
$R^1 = R^2 = Ph; 64\%$
$R^1 = R^2 = Me; 0\%$ 23%
$R^1 = Me, R^2 = Ph; 25\%$ 29%

$R^1 = R^2 = H; 72\%$
$R^1 = Me, R^2 = H; 65\%$
$R^1 = H, R^2 = Me; 40\%$

40%

Scheme 56

3.1.7
[4+6] Cycloaddition

A variety of thiophene dioxides including the parent compound 1 undergo a [4+6] cycloaddition with 6-(dimethylamino)fulvene and its 6-methyl and 6-dimethylamino derivatives to provide a facile synthesis of azurene and its derivatives **95**, though the yields are generally low (Scheme 57) [177–182]. The reaction affords an isomeric mixture when unsymmetrically substituted thiophene dioxides are used as the starting material. The probable initial adducts **94** spontaneously extrude dimethylamine and sulfur dioxide to produce **95**. Benzocyclopropene also undergoes [4+6] cycloaddition with thiophene dioxides to result in a one-pot formation of methanonaphthalenes **97** in modest yields [183]. In this case, extrusion of sulfur dioxide and ring opening of the initial adducts **96** take place simultaneously (Scheme 58).

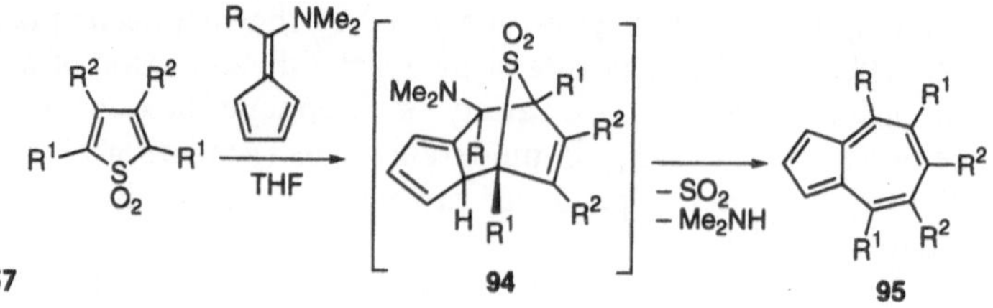

Scheme 57

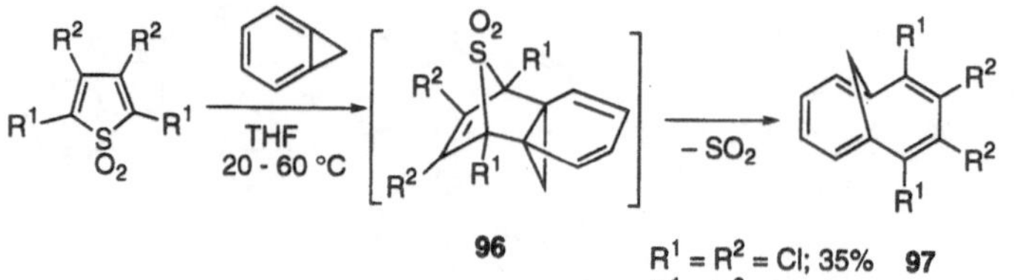

Scheme 58

$R^1 = R^2 = Cl; 35\%$ **97**
$R^1 = R^2 = Br; 25\%$
$R^1 = Me, R^2 = Br; 5\%$
$R^1 = H, R^2 = Ph; 5\%$

3.1.8
Ene Reaction

Only one example of an ene reaction is reported [40]. As already described, benzyne and thiophene dioxides undergo [4+2] cycloaddition. However, the reaction of 3,4-dineopentylthiophene 1,1-dioxide with benzyne affords compound **98**, the product of an ene reaction. The other two products, **99** and **100**, result from [4+2] cycloaddition (Scheme 59).

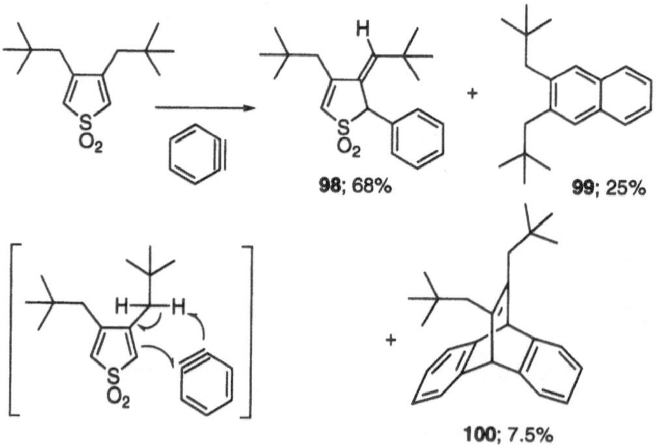

98; 68%   **99; 25%**

Scheme 59   **100; 7.5%**

3.2
Nucleophilic Addition

Thiophene 1,1-dioxides are typical α,β-unsaturated sulfones and take part in a nucleophilic 1,4-addition acting as Michael acceptors. Thus, a wide variety of nucleophiles add to the double bond of thiophene 1,1-dioxides. If a good leaving-group such as a halogen exists on the double bond to which a nucleophile adds, then an addition-elimination may take place to give substituted thiophene 1,1-dioxides.

Ammonia [184], benzylamine [55, 58], dimethylamine [57], and thiophenolate [59] add to the parent thiophene 1,1-dioxide 1. If nucleophiles are used in excess, bis-adducts will form. The formation of these adducts has been used as evidence for the generation of 1 (Scheme 60).

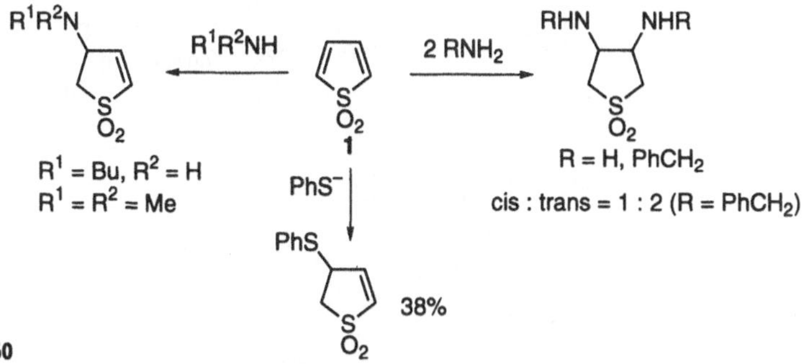

$R^1 = Bu, R^2 = H$
$R^1 = R^2 = Me$

$R = H, PhCH_2$

cis : trans = 1 : 2 (R = PhCH$_2$)

38%

Scheme 60

The addition of α-toluenethiolate to 3,4-dimethylthiophene 1,1-dioxide 101 affords the sulfolene 102 (Scheme 61) [185]. The previously reported structure 103 is erroneous [186]. The dioxide 101 is in an equilibrium with its isomer 7 under the basic conditions. The addition of the thiolate to 7 would give rise to 102. The reaction of α-toluenethiolate with 2,5-dimethylthiophene 1,1-dioxide 104 yields compound 105 by isomerization of the initial Michael adduct (Scheme 62).

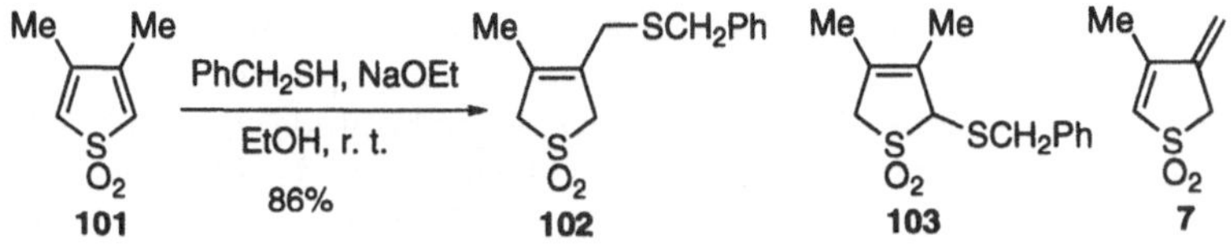

Scheme 61

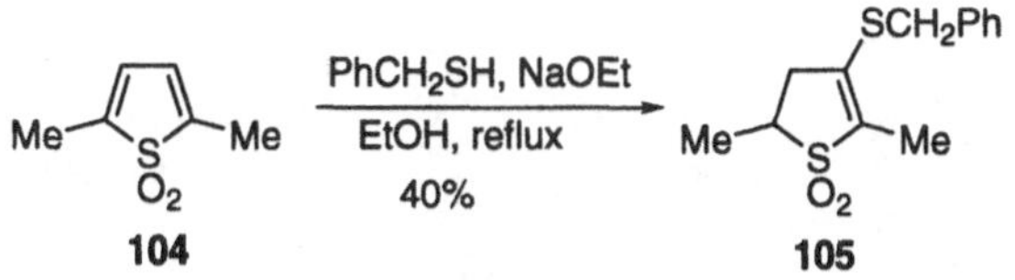

Scheme 62

The reaction of the dioxide 106 with α-toluenethiolate in DMF affords a substitution product 107 by an addition-elimination mechanism (Scheme 63) [185]. Meanwhile, the reaction in refluxing ethanol gives a mixture of compounds 108 and 109 through addition of the thiolate to two exo-methylene isomers of 106, followed by reaction with ethanol.

Pyrrolidine and piperidine add in the usual manner to the less hindered double bond of thiophene 1,1-dioxides (Scheme 64) [187, 188].

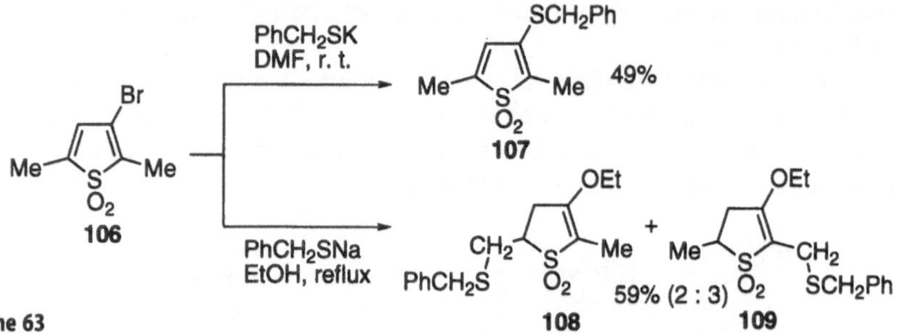

Scheme 63

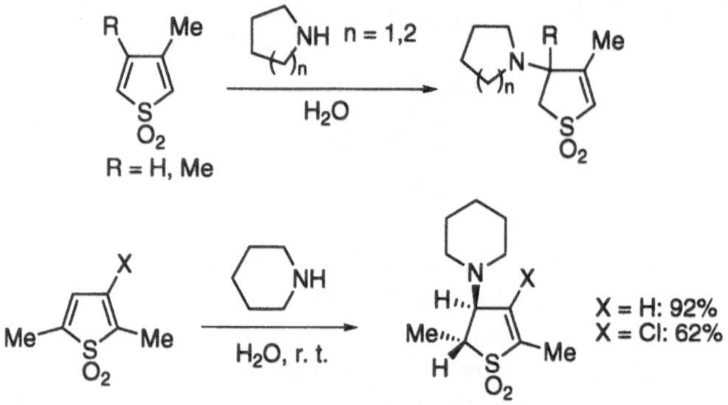

Scheme 64

Benzo[b]thiophene 1,1-dioxide 38 and its derivatives act as typical α, β-unsaturated sulfones, and a variety of nucleophiles add to the carbon atom at the 3-position. Thus, amines and the conjugated bases of water, alcohols, thiophenol, and diethyl malonate all add to 38 to give the corresponding adducts 110 in excellent yields (Scheme 65) [189]. Even hydrogen bromide can add to 38. When 2-bromobenzo[b]thiophene 1,1-dioxide 111 was treated with piperidine and morpholine under appropriate conditions, excellent yields of adducts 112 were isolated without being accompanied by dehydrobromination (Scheme 66) [190].

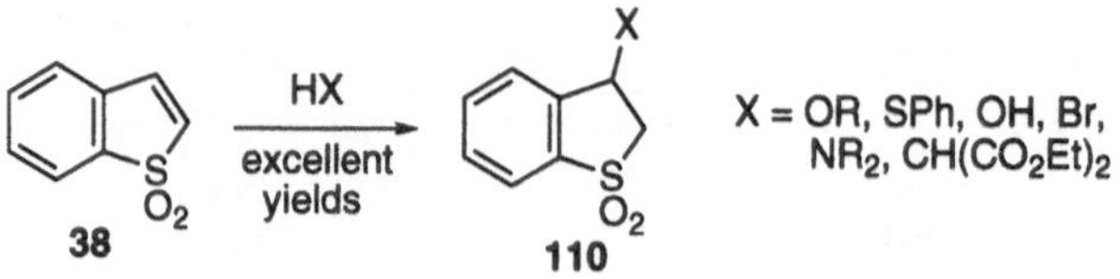

Scheme 65

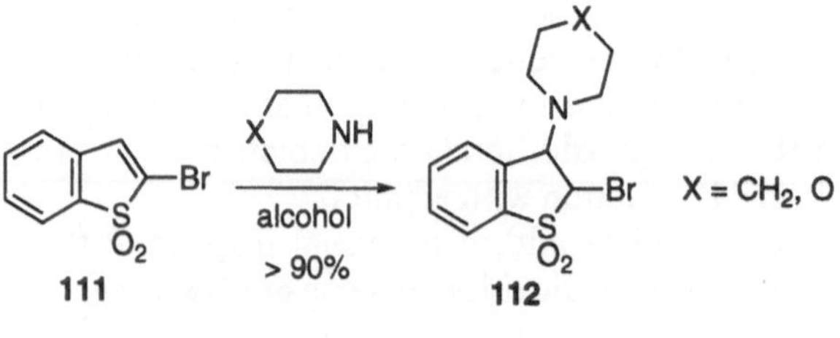

Scheme 66

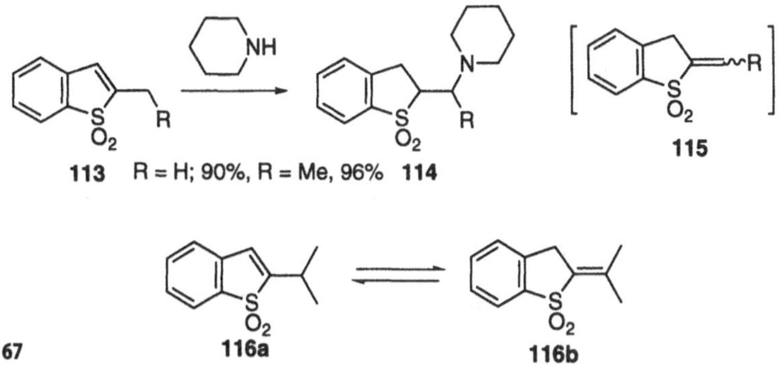

NH
R
R
O_2
113 R = H; 90%, R = Me, 96% 114
R
115
116a
116b

Scheme 67

On reaction with piperidine, 2-alkylbenzo[*b*]thiophene 1,1-dioxides **113** afford compounds **114** in excellent yields, probably because the addition proceeds through the exocyclic alkene **115** (Scheme 67) [191]. In fact, the dioxide **116a**, on heating with piperidine, affords an equilibrium mixture with its exocyclic isomer **116b**, where the latter is the predominant component.

3-Bromobenzo[*b*]thiophene 1,1-dioxide **117** and 3-bromo-2-phenylbenzo[*b*]thiophene 1,1-dioxide **119** reacted with ammonia and primary and secondary amines to give the corresponding 3-aminobenzothiophene dioxides, **118** and **120**, in good yields (Scheme 68) [17, 192–194]. 2-Benzoyl-3-chlorobenzo[*b*]thiophene 1,1-dioxide **121** also reacted with a series of nitrogen nucleophiles to give substitution products **122** in good yields (Scheme 69) [195].

Reactions of 2,3-dichlorobenzo[*b*]thiophene 1,1-dioxide **123** with nitrogen and oxygen nucleophiles produce substitution products, **124** and **125**, at the 3-position because the addition of nucleophiles takes place exclusively at that position (Scheme 70) [196–198]. However, in the case of 2-bromo-3-phenylbenzo[*b*]thiophene 1,1-dioxide **126**, nitrogen nucleophiles added to carbon 2 to give compounds **127** (Scheme 71) [194].

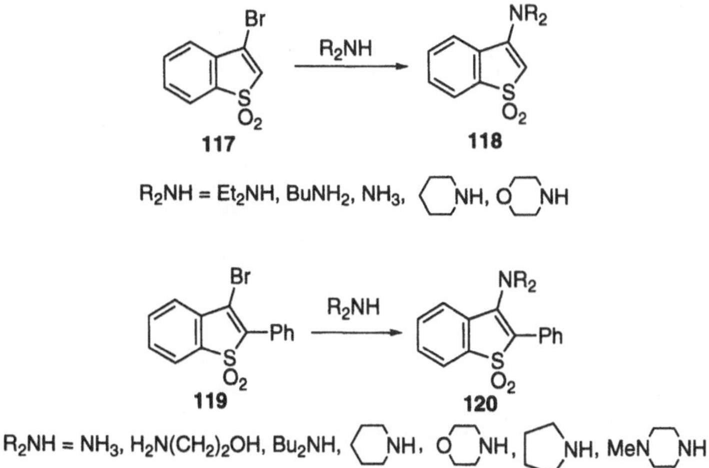

Br
R_2NH
NR_2
117
118
R_2NH = Et_2NH, $BuNH_2$, NH_3, NH, O NH
Br
Ph
R_2NH
NR_2
Ph
119
120
R_2NH = NH_3, $H_2N(CH_2)_2OH$, Bu_2NH, NH, O NH, NH, MeN NH

Scheme 68

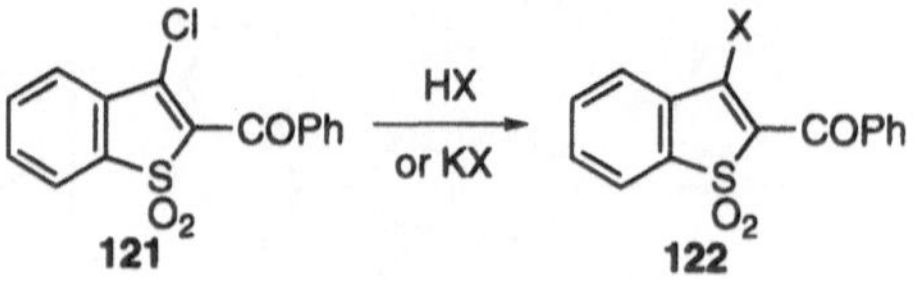

X	Yields (%)
-NHNHCONH$_2$	50
-NHN=C(NH$_2$)Tol	60
-NH$_2$	80
-NCS	90
-HN⟨⟩	40
H$_2$N⟨⟩	

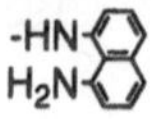

Scheme 69

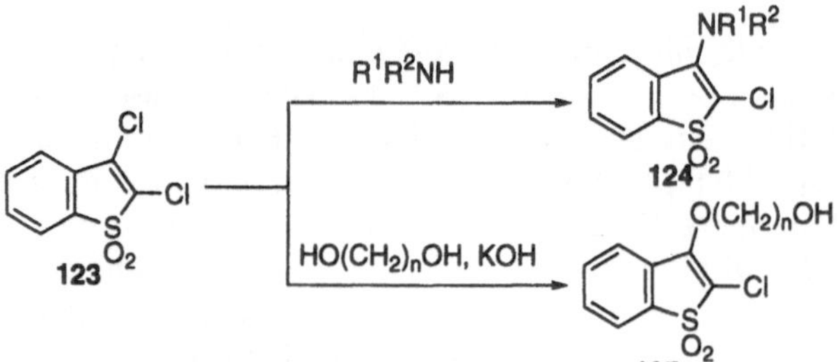

Scheme 70

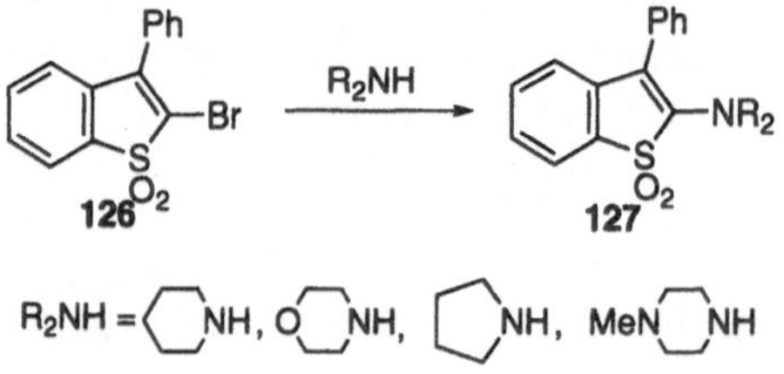

Scheme 71

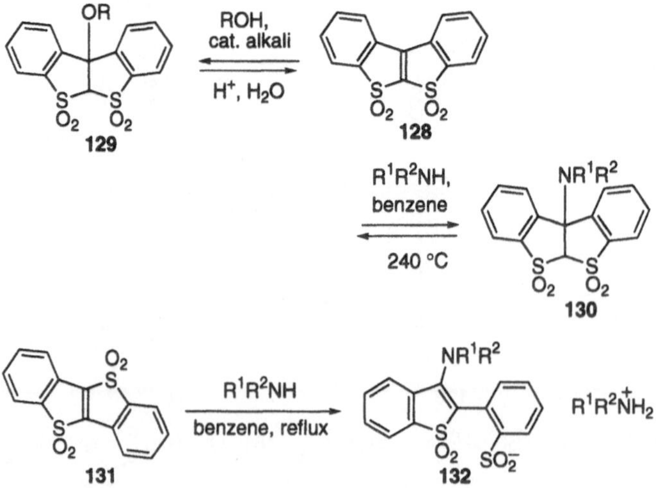

$$R_2NH = \text{⟨}NH, \text{ O⟨}NH, \text{ ⟨}NH, \text{ MeN⟨}NH$$

Scheme 72

Addition of oxygen and nitrogen nucleophiles to the dioxide **128** yielding **129** and **130**, respectively, is reversible [199], while amine adducts of the dioxide **131** undergo ring opening to give sulfinic acid derivatives **132** (Scheme 72) [200].

Ring opening of the primary adducts is often encountered. For example, the adducts **134** of the dioxides **133** with morpholine and piperidine further react with these amines to give amides **135** in good yields (Scheme 73) [201]. The reaction of **136** with hydrazine under forcing conditions affords the sulfone **137** in 77 % yield (Scheme 74) [202]. The ring opening reaction of thiophene 1,1-dioxides is discussed later.

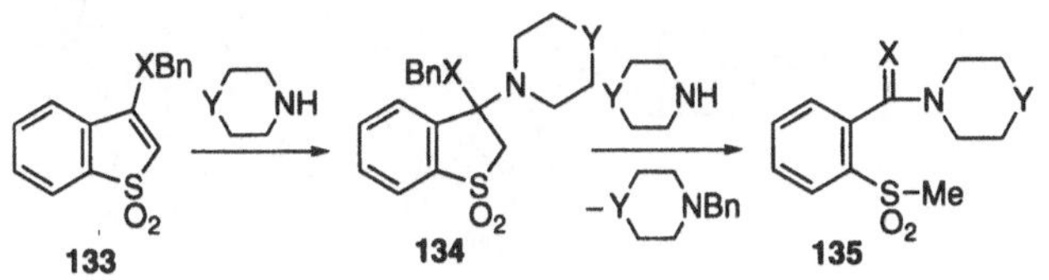

Scheme 73   X = O, Y = CH$_2$; 97%: X = O, Y = O; 86%: X = S, Y = O; 95%

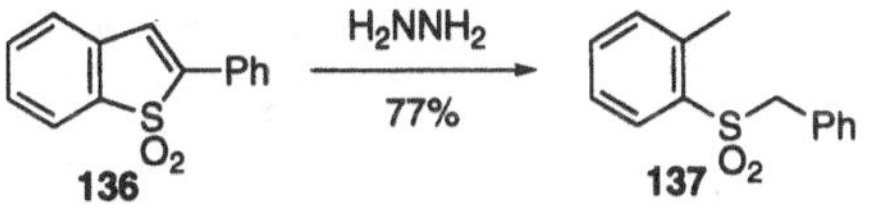

Scheme 74   **136**   **137**

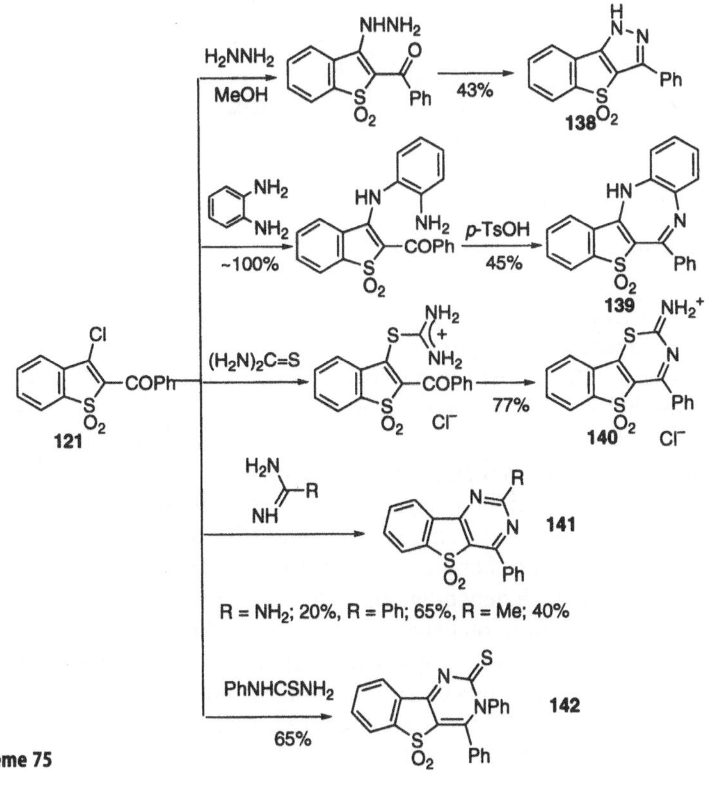

Scheme 75

Michael addition of appropriate nucleophiles to benzo[b]thiophene 1,1-dioxides followed by intramolecular condensation leads to the formation of heterocycle-annelated benzothiophene 1,1-dioxides. Thus, a wide variety of heterocycle-annelated compounds such as **138–142** were synthesized by addition of nucleophiles to 2-benzoyl-3-chlorobenzo[b]thiophene 1,1-dioxide **121** (Scheme 75) [195, 203, 204].

Other examples of heterocycle-annelation using a carboxylic acid or an acid chloride derivative of benzo[b]thiophene dioxides are given below (Scheme 76) [205–207].

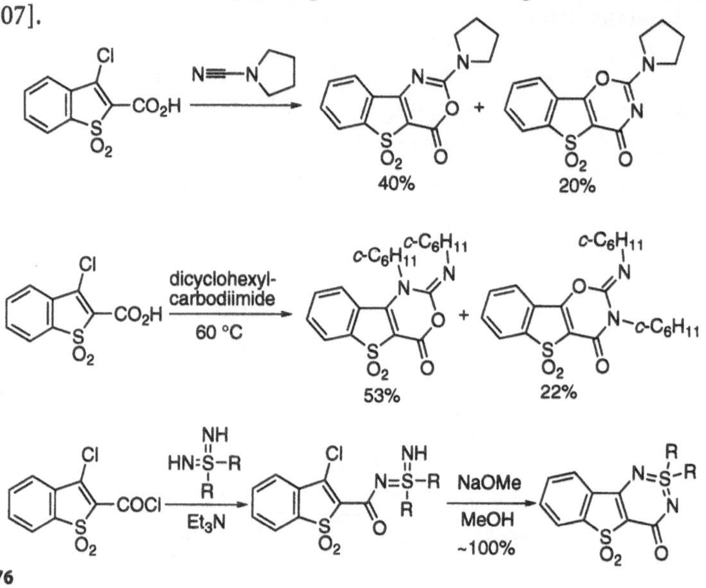

Scheme 76

3.3
Ring Opening Reactions

The synthetic utility of ring opening reactions of thiophenes and thiophene 1,1-dioxides was reviewed recently [208].

When a series of 2,5-disubstituted 3-bromothiophene 1,1-dioxides **143** were treated with methyl, ethyl, butyl, *tert*-butyl, and phenyllithiums, they were consumed in two competing ring-opening reactions (paths a and b) (Scheme 77). In path a, bromine-lithium exchange followed by ring opening produces enesulfinates **144**, which are trapped by benzyl bromide to give sulfones **145**. In path b, 1,6-Michael addition followed by ring opening and elimination of SO_2 and LiBr produces isomeric mixtures of enynes **146a** and **146b**. The ratio of the two ring opening reactions was much dependent upon the bulkiness of the 5-substituent, but also on the organolithium reagent used [209, 210]. The corresponding chloro derivatives such as 3-chloro-2,5-dimethylthiophene 1,1-dioxide gave only 1,6-Michael addition product and no chlorine-lithium exchange product. In the case of the dioxide **147**, bromine-lithium exchange reaction resulted in selective formation of the sulfone **148** in 31% yield (Scheme 78) [211, 212].

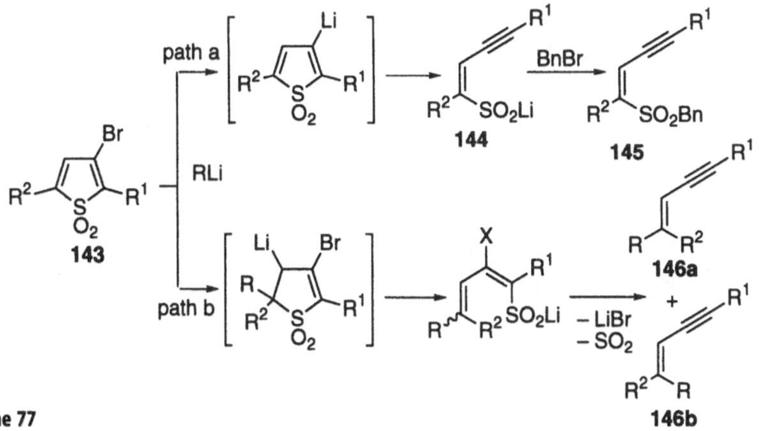

Scheme 77

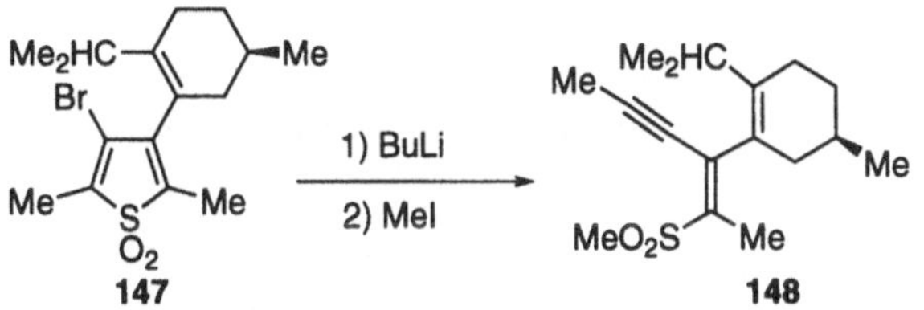

Scheme 78

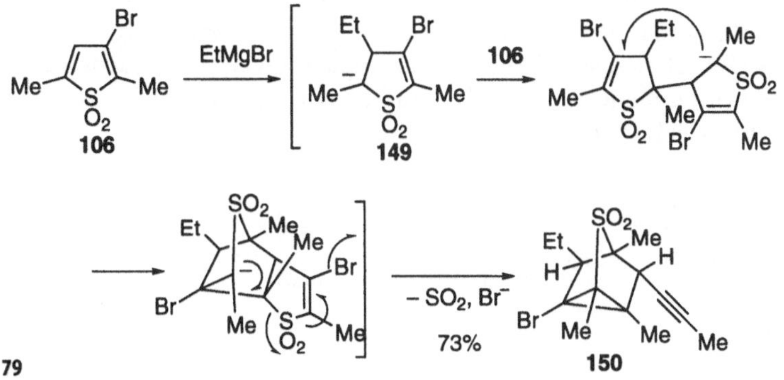

Scheme 79

In contrast to organolithium reagents, Grignard reagents such as EtMgBr reacted with 3-bromo-2,5-dimethylthiophene 1,1-dioxide 106 in a 1,4-Michael addition manner. The resulting adduct 149 further reacted with 106 to give the cage compound 150 as the final product in 73% yield (Scheme 79). PrMgBr reacted similarly, but no simple products were obtained from MeMgBr or PhMgBr.

The dioxide 106 reacts with cyclic secondary amines in refluxing toluene to give 2,4-hexadiene derivatives 153 (Scheme 80) [188]. Addition of the amines to the exo-methylene isomer 151, present in a minute amount in the equilibrium, is probably responsible for the formation of 153. The corresponding chloro derivative also reacted with these amines in a similar manner. Application of this

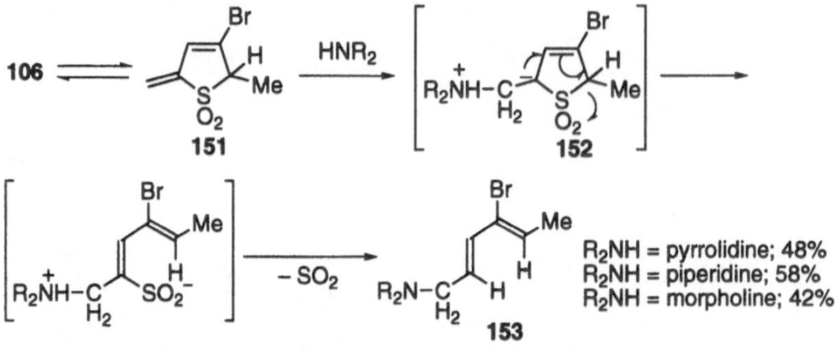

Scheme 80

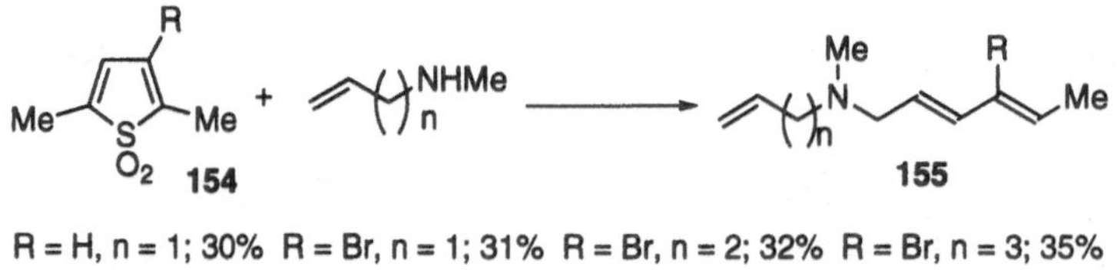

R = H, n = 1; 30% R = Br, n = 1; 31% R = Br, n = 2; 32% R = Br, n = 3; 35%

Scheme 81

reaction to **154** and ω-unsaturated amines provides a new route to azatrienes **155** (Scheme 81) [213].

As described in Sect. 3.1.2, some thiophene dioxides dimerize in a [2+4] cycloaddition manner and the resulting adducts undergo ring opening to give alkynes [105, 107]. A supplementary example is given below (Scheme 82).

Dibenzothiophene 5,5-dioxide undergoes ring opening by action of KOH [214] and ROK (Scheme 83) [215].

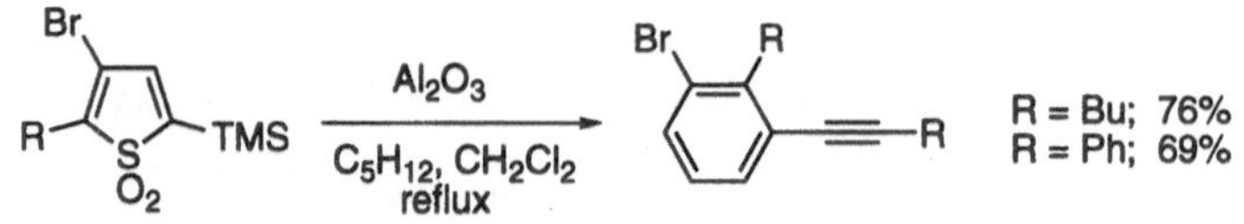

Scheme 82

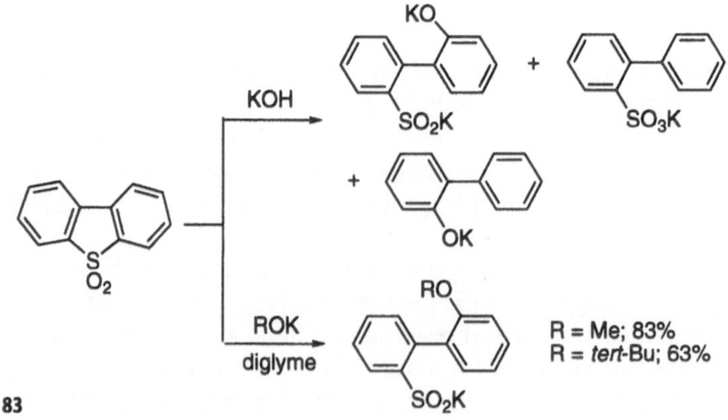

Scheme 83

3.4
Ring Alkylation

Ring alkylation of thiophene 1,1-dioxides was performed in two ways. Treatment of 3-bromo-2,5-dimethylthiophene 1,1-dioxide 106 with alkylcopper (RCu) or lithium dialkylcuprate (R_2CuLi) affords 3-alkyl-2,5-dimethylthiophene 1,1-dioxides 154 in good yields (Scheme 84) [216]. The use of the former reagent afforded better yields of 154 without side-reactions. The reaction involves Michael addition and debromination. Although 3,4-di-*tert*-butyl- and 3,4-di-(1-adamantyl)thiophenes are not lithiated by alkyllithiums and LDA because of steric hindrance, their dioxide derivatives 87 are more acidic and easily lithiated by LDA. Treatment of the resulting lithiated species with methyl iodide produced the corresponding highly congested thiophene 1,1-dioxides 156 in good yields (Scheme 85) [217].

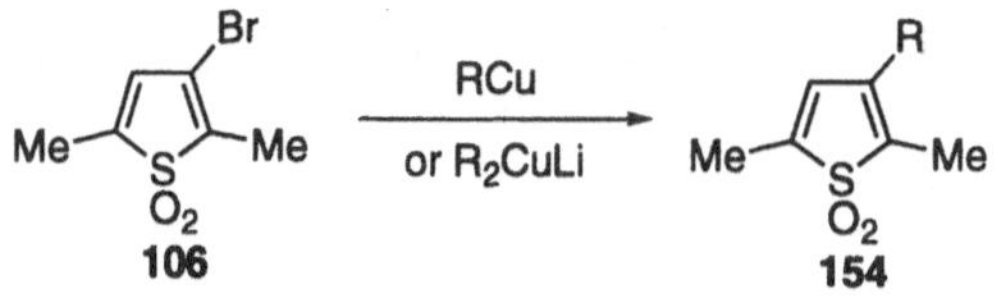

RCu: R = Me, 100%; R = Et, 100%; R = Bu, 100%; R = *tert*-Bu; 47%
R_2CuLi: R = Me, 78%; R = Et, 83%; R = Bu, 53%; R = *tert*-Bu; 11%

Scheme 84

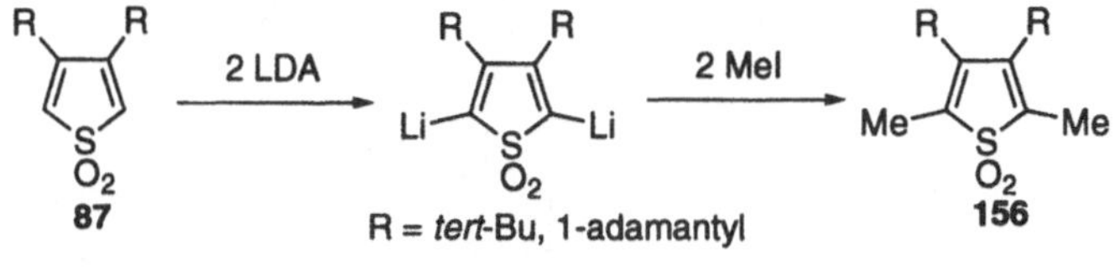

R = *tert*-Bu, 1-adamantyl

Scheme 85

3.5
S_N2' Reactions

2-Chloromethylbenzo[*b*]thiophene 1,1-dioxide 157 reacts with piperidine, thiourea, and sodium thiophenoxide in an S_N2 mechanism to give usual substitution products 158 in reasonable yields (Scheme 86). Meanwhile, 3-chloromethylbenzo[*b*]thiophene 1,1-dioxide 159 reacts with piperidine, morpholine, thiourea, and sodium thiophenoxide in an S_N2' mode (see the mechanism shown in brackets) to give thiophene dioxides 160 (Scheme 87). Application of the reaction to dioxides 161 affords rearranged compounds 163 in good yields because of ring opening, internal rotation, and ring closure of the S_N2' products 162 (Scheme 88). This type of reactions have been investigated in great detail from a mechanistic point of view [218–225].

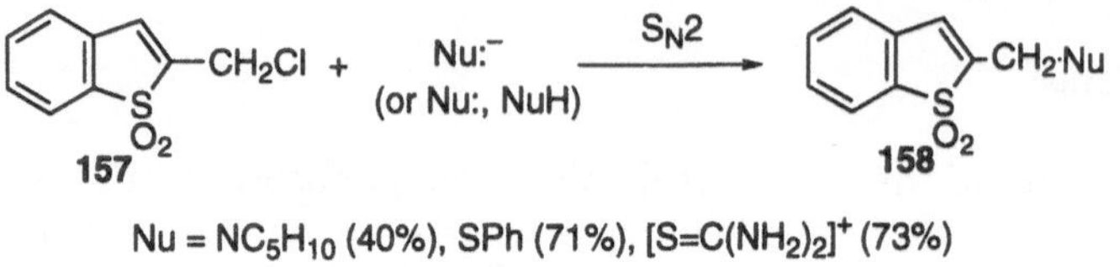
157
Nu:⁻
(or Nu:, NuH)
SN2
158
Nu = NC5H10 (40%), SPh (71%), [S=C(NH2)2]+ (73%)

Scheme 86

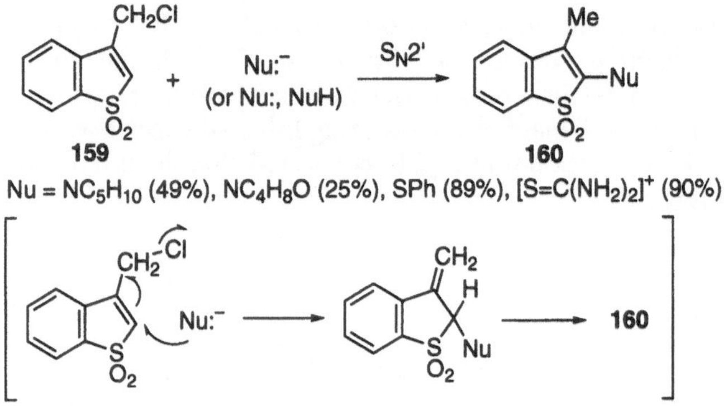
CH2Cl
Nu:⁻
(or Nu:, NuH)
SN2'
Me
Nu
159
160
Nu = NC5H10 (49%), NC4H8O (25%), SPh (89%), [S=C(NH2)2]+ (90%)
160

Scheme 87

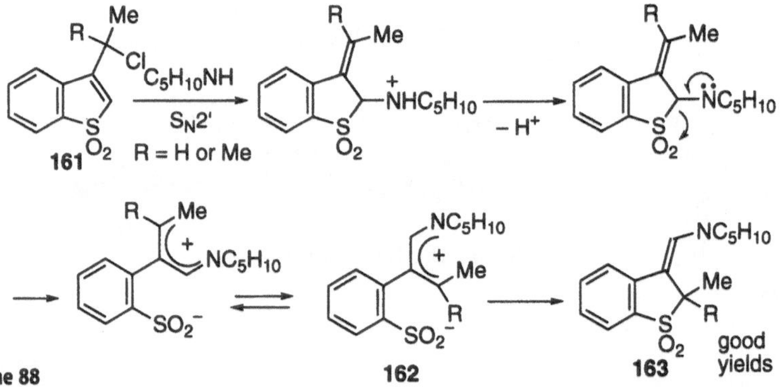
C5H10NH
SN2'
161
R = H or Me
NHC5H10
- H+
NC5H10
162
163
good
yields

Scheme 88

3.6
Steglich Reagent

The cyclic carbonate derivative **11a** of 3,4-dihydroxy-2,5-diphenylthiophene 1,1-dioxide **9**, Steglich reagent, is a strongly activating reagent in many condensation reactions. It is now commercially available, though easily prepared by treatment of **9** with phosgene. Treatment of carboxylic acids with **11a**, treatment of carboxylic acids with **11a** in the presence of dicyclohexylcarbodiimide, and treatment of acid chlorides with **11a** all afforded the esters **164** in good yields, which were converted into the dipeptides **165** in excellent yields on reactions with amines (Scheme 89) [84].

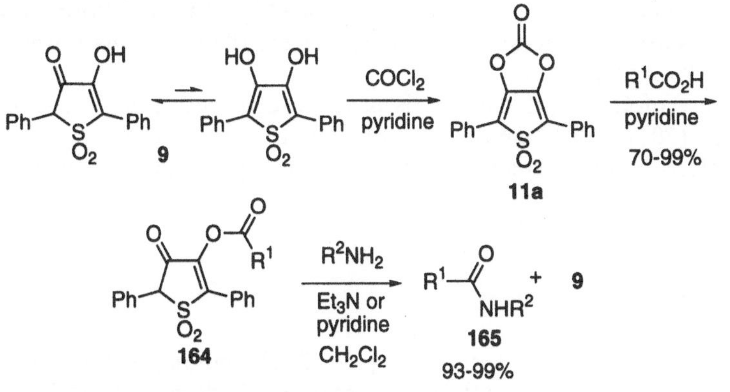

R^{1}CO = EtCO, *t*-BuCO, Cbz-Val, Cbz-Pro, Boc-Pro, Boc-Phe, Boc-Met
R^{2}NH$_2$ = Val-OMe, Val-O-*t*-Bu

Scheme 89

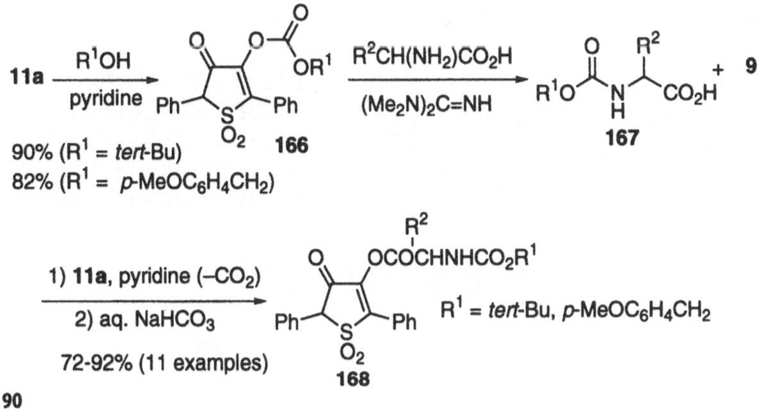

Scheme 90

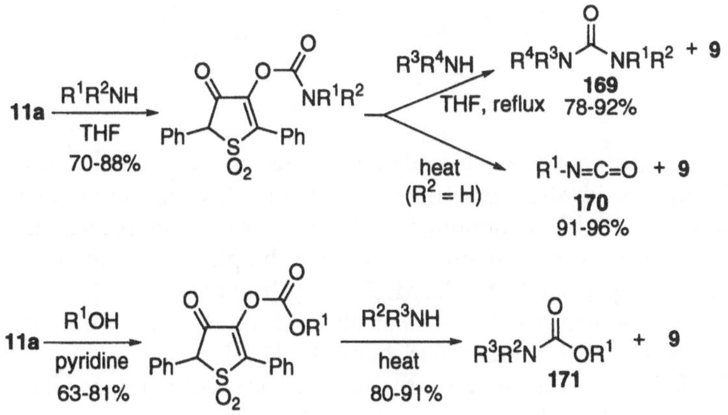

Scheme 91

Treatment of **11a** with *tert*-BuOH and *p*-MeOC$_6$H$_4$CH$_2$OH affords carbonates **166**, which are converted into amino acid esters **168** in excellent overall yields (Scheme 90) [226].

The use of **11a** for the synthesis of di- and tri-substituted ureas **169**, iso-cyanates **170**, and urethanes **171** were reported (Scheme 91) [227]. Further applications to the preparation of amides and peptides have also appeared [228–231].

3.7
Reduction

Monocyclic thiophene 1,1-dioxides [186, 232] and benzo[*b*]thiophene 1,1-dioxide [233] are reduced to the corresponding thiophenes by zinc in a mixture of acetic acid and hydrochloric acid (Scheme 92). Dibenzothiophene 5,5-dioxide is reduced to dibenzothiophene in many ways (Scheme 93): BuLi/NaH in ether/THF [234], NaH/*tert*-AmONa/Ni(OAc)$_2$/2,2′-bipyridyl in DME [235], LiAlH$_4$/nickelocene in THF [236], and S$_8$ (320 °C) [237] are the reagents used for this conversion.

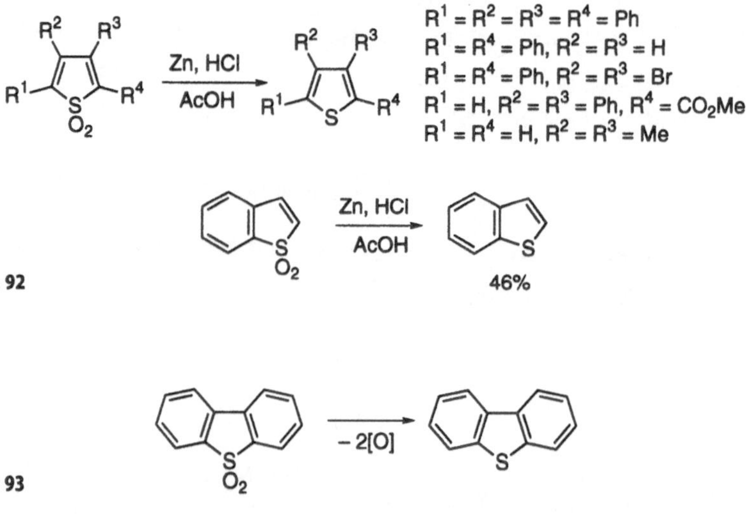

$R^1 = R^2 = R^3 = R^4 = Ph$
$R^1 = R^4 = Ph, R^2 = R^3 = H$
$R^1 = R^4 = Ph, R^2 = R^3 = Br$
$R^1 = H, R^2 = R^3 = Ph, R^4 = CO_2Me$
$R^1 = R^4 = H, R^2 = R^3 = Me$

Scheme 92

46%

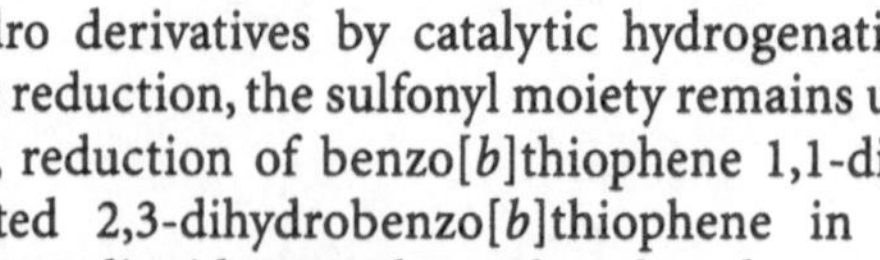

Scheme 93

Benzo[*b*]thiophene 1,1-dioxide and its derivatives are reduced to the corresponding 2,3-dihydro derivatives by catalytic hydrogenation in good yields (Scheme 94). In this reduction, the sulfonyl moiety remains unchanged [13, 189, 238, 239]. However, reduction of benzo[*b*]thiophene 1,1-dioxide with LiAlH$_4$ affords deoxygenated 2,3-dihydrobenzo[*b*]thiophene in 79% yield [233]. Benzo[*b*]thiophene 1,1-dioxide was also reduced to the 2,3-dihydro derivative by an electrochemical method (Scheme 95) [240].

The followings are some synthetically important reactions where reductive process is involved (Scheme 96) [241–243].

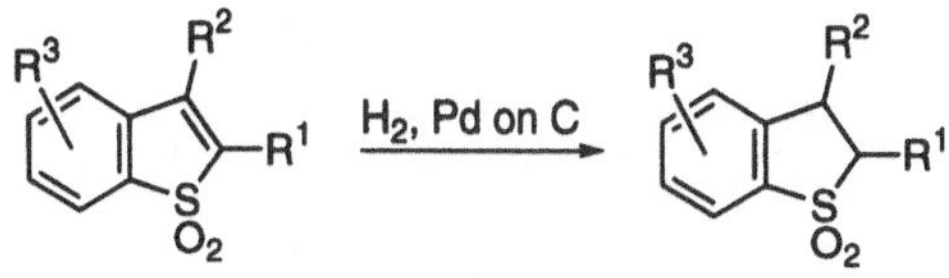

Scheme 94

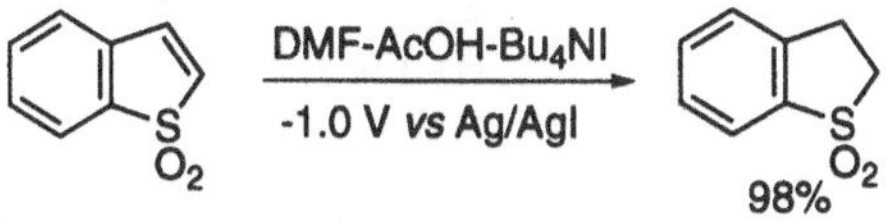

Scheme 95

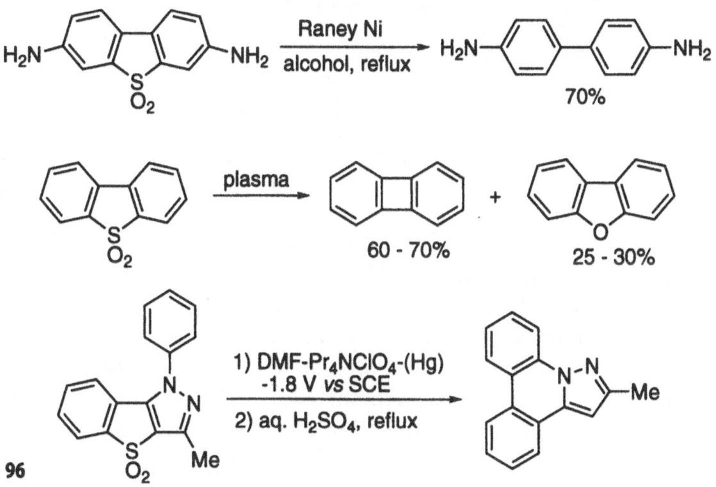

Scheme 96

3.8
Oxidation

Oxidation of benzo[b]thiophene 1,1-dioxide with $KMnO_4$ affords the sulfonic acid **172** [189], while oxidation with alkaline H_2O_2 yields the ketone **173** (Scheme 97) [244]. Oxidation of 3-substituted benzo[b]thiophene 1,1-dioxides **174** with alkaline H_2O_2 affords the alcohols **175** as the principal product (Scheme 98) [244]. Compounds **173** and **175** are presumably formed through Michael addition of HOO^- to the 3-position of the substrates. Indeed, the reaction of 3,4-di-tert-butylthiophene 1,1-dioxide **83** with H_2O_2 under alkaline conditions at room temperature afforded the Michael adduct **177** in 91% yield, while the reaction carried out at 50–60 °C gave the ring-opened product **178** in 15% yield in addi-

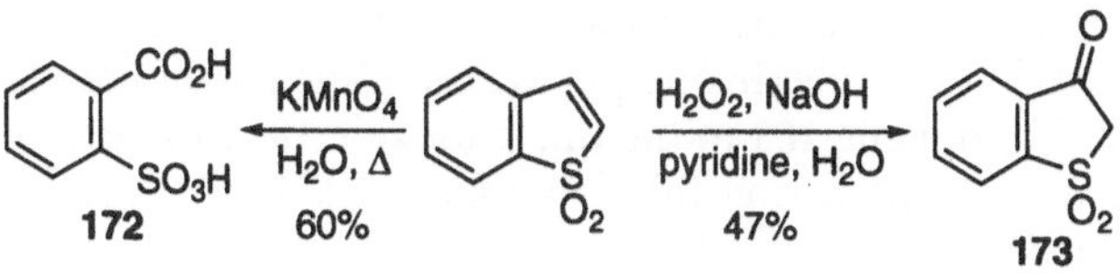

Scheme 97

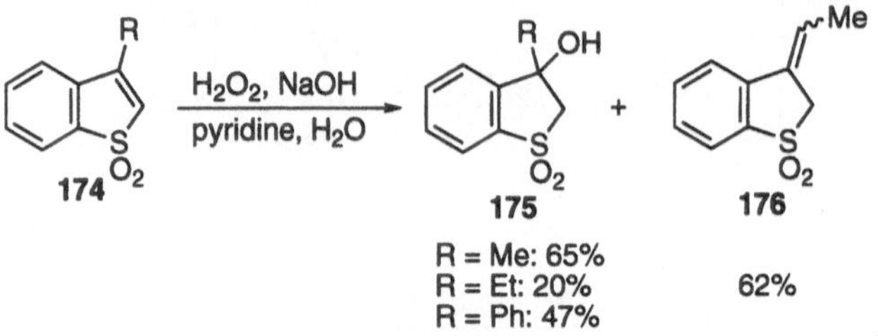

Scheme 98

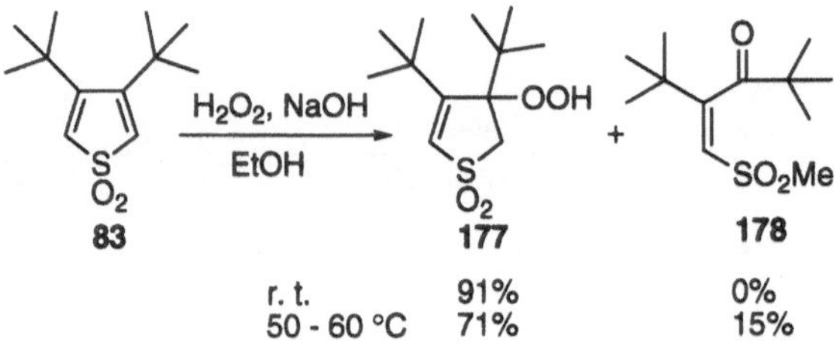

R = Me: 65%
R = Et: 20%
R = Ph: 47%

62%

Scheme 99

	177	178
r. t.	91%	0%
50 - 60 °C	71%	15%

tion to **177** (Scheme 99) [245]. Compound **178** is produced by base-induced decomposition of **177**; treatment of **177** with sodium hydroxide in refluxing ethanol afforded **178** in 88% yield.

Oxidation of thiophene dioxides **6** with m-CPBA in the presence of sodium carbonate proceeds rather slowly, but affords the corresponding epoxides, **179** and **180**, generally in good yields (Scheme 100) [217, 245, 246]. The oxidation takes place more quickly with highly congested substrates such as 3,4-di-$tert$-butyl-2,5-dimethyl- and 3,4-di(1-adamantyl)-2,5-dimethylthiophene 1,1-dioxides (Table 11). This is probably due to the activation of the double bonds by steric hindrance (destabilization of HOMO) and also relief of steric crowding for the epoxide formation.

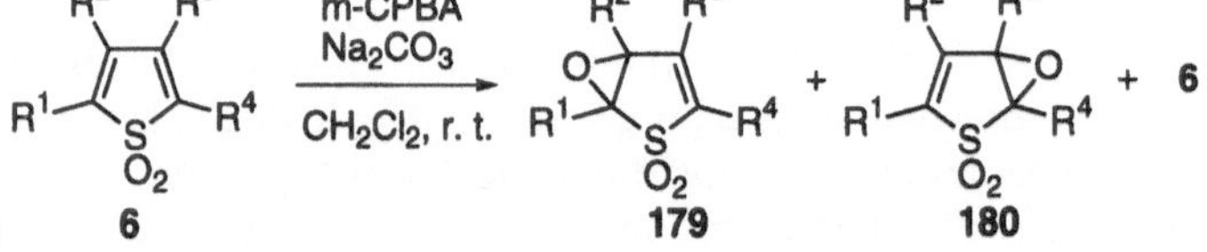

Scheme 100

However, the oxidation of **6** carried out in the absence of sodium carbonate affords thiete 1,1-dioxides, **181** and **182**, when highly congested substrates were used (Scheme 101, Table 12) [217, 246]. The thiete dioxides are probably formed by acid-catalyzed rearrangement of the epoxides, although their direct formation from **6** cannot be ruled out. The reaction provides an interesting synthesis of thiete dioxides. In particular, oxidation of congested thiophenes may afford the corresponding thiete dioxides in one-pot as exemplified in the conversion of 3,4-di(1-adamantyl)thiophene **183** to the thiete dioxide **184** in 78% yield (Scheme 102). Oxidation of a tropone ring-fused thiophene 1,1-dioxide with m-CPBA was also reported [247].

Table 11. Oxidation of thiophene 1,1-dioxides (**6**) with *m*-CPBA under basic conditions

R^1	R^2	R^3	R^4	Reaction time	Yield (%)		
					179	**180**	**6**
H	*t*-Bu	*t*-Bu	H	21 d[a]	82		
H	1-ad	1-ad	H	8 d	71		10
Me	1-ad	1-ad	Me	21 h	99		
Me	*t*-Bu	*t*-Bu	Me	23 h	85		
Me	*t*-Bu	*t*-Bu	H	14 d	17	50	
Me	*t*-Bu	H	*t*-Bu	7 d	33		65
Me	*t*-Bu	H	Me	7 d	19	4	58
Me	Me	Me	Me	48 h	40		20

[a] Reflux in $Cl(CH_2)_2Cl$.

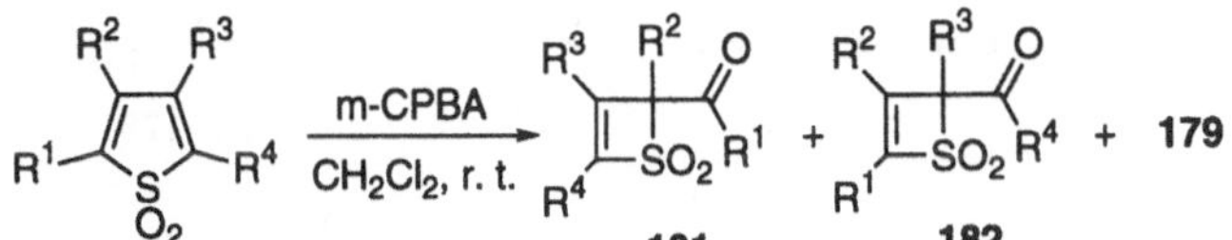

Scheme 101 **6** **181** **182**

Table 12. Oxidation of Thiophene 1,1-Dioxides (**6**) with *m*-CPBA

R^1	R^2	R^3	R^4	Reaction time	Yield (%)		
					181	**182**	**179**
H	*t*-Bu	*t*-Bu	H	14 d	3		79
H	1-ad	1-ad	H	9 d	trace		78
Me	1-ad	1-ad	Me	40 h	97		
Me	*t*-Bu	*t*-Bu	Me	19 h	95		
Me	*t*-Bu	*t*-Bu	H	19 d	19	44	
Me	*t*-Bu	H	*t*-Bu	7 d		4	

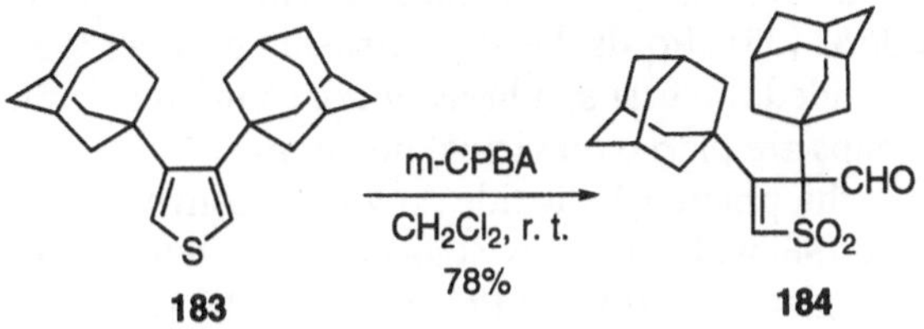

Scheme 102 **183** **184**

3.9
Pyrolysis

Thermal decomposition of thiophene 1,1-dioxides has already been partly described in connection with their dimerization (Sect. 3.1.2). Pyrolysis products of thiophene 1,1-dioxides are much dependent on reaction conditions. Thus, pyrolysis of tetraphenylthiophene 1,1-dioxide 185 produced compounds 186–192 depending upon the reaction conditions (Scheme 103) [248].

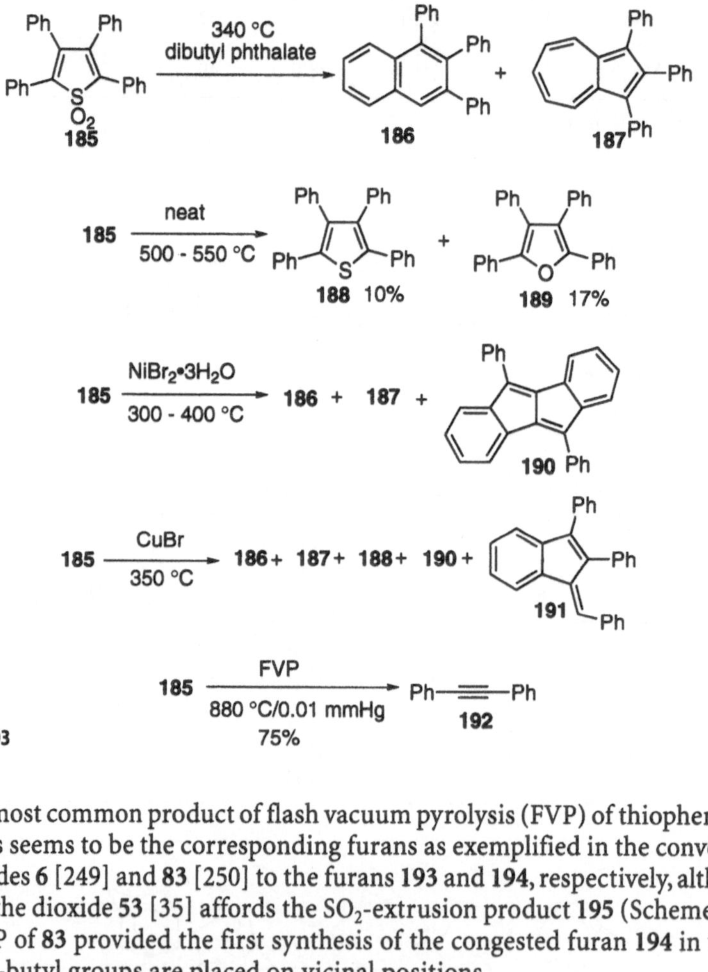

Scheme 103

The most common product of flash vacuum pyrolysis (FVP) of thiophene 1,1-dioxides seems to be the corresponding furans as exemplified in the conversion of dioxides 6 [249] and 83 [250] to the furans 193 and 194, respectively, although FVP of the dioxide 53 [35] affords the SO$_2$-extrusion product 195 (Scheme 104). The FVP of 83 provided the first synthesis of the congested furan 194 in which two *tert*-butyl groups are placed on vicinal positions.

FVP of benzo[b]thiophene 1,1-dioxide provides a convenient synthesis of the parent benzothiete 196, which is fairly stable at room temperature, but dimerizes above 100 °C to give the dithiocin 197 (Scheme 105) [251].

Pyrolysis of dibenzothiophene 5,5-dioxide and the derivatives affords mixtures of the corresponding dibenzofurans and dibenzothiophenes (Scheme 106) [214, 252, 253].

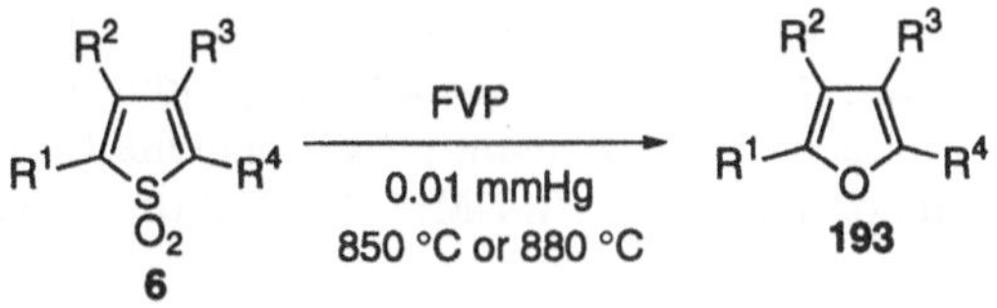

$R^1 = R^4 = $ Me, $R^2 = R^3 = $ H; 67%, $R^1 = R^4 = $ t-Bu, $R^2 = R^3 = $ H; 47%, $R^1 = R^3 = $ t-Bu, $R^2 = R^4 = $ H; 48%, $R^1 = R^4 = $ Ph, $R^2 = R^3 = $ H; 53%

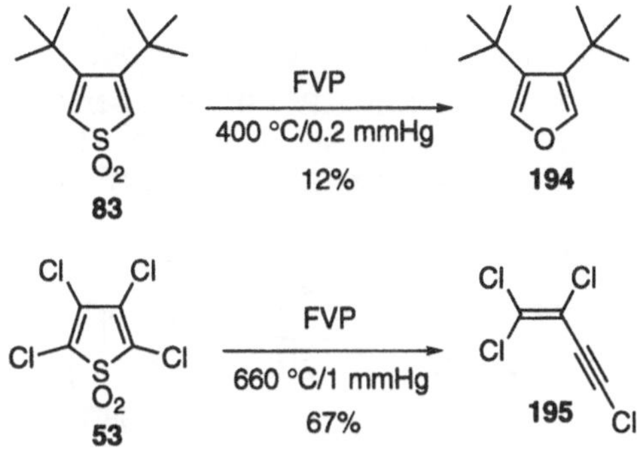

Scheme 104

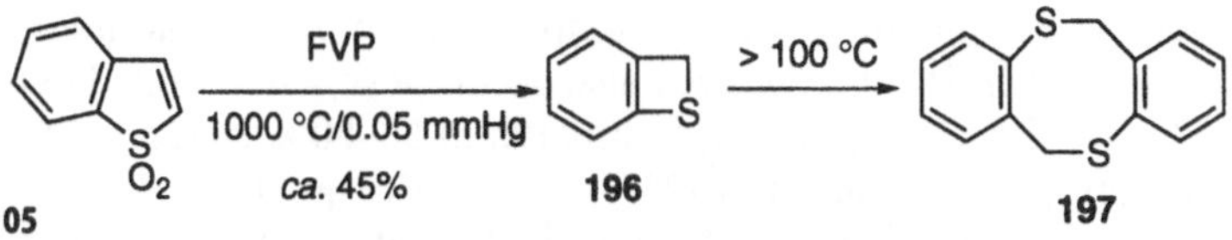

Scheme 105

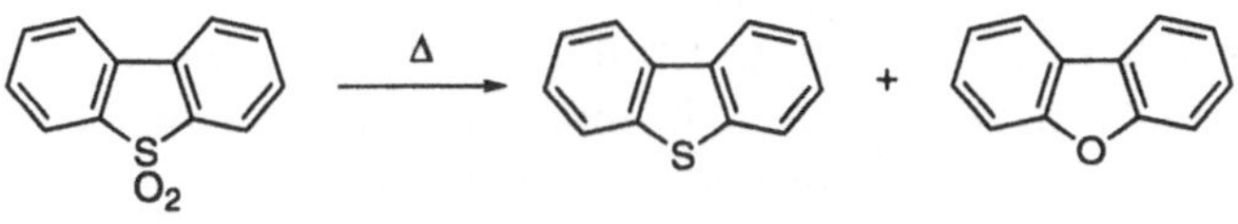

Scheme 106

4
Selenophene 1,1-Dioxides

Although the chemistry of selenophene 1,1-dioxides seems equally fruitful, it has been much less studied. It is only recently that the first successful synthesis of selenophene 1,1-dioxides appeared [254, 255]. Oxidation of tetraarylseleno-phenes **198** with m-CPBA did not afford the corresponding selenophene 1,1-di-oxides. The reaction produced E-1,2-diaryl-1,2-diaroylethylenes **199** as the prin-

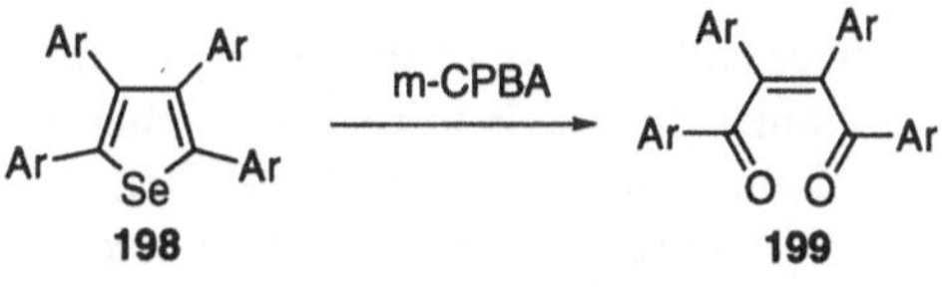

Scheme 107

cipal product (Scheme 107) [256]. On the other hand, oxidation of selenophenes **200** with dimethyldioxirane (DMD) satisfactorily produced the corresponding selenophene 1,1-dioxides **201** in excellent yields (Scheme 108). Similarly, oxidation of benzo[*b*]selenophene **202** with DMD afforded the dioxide **203**.

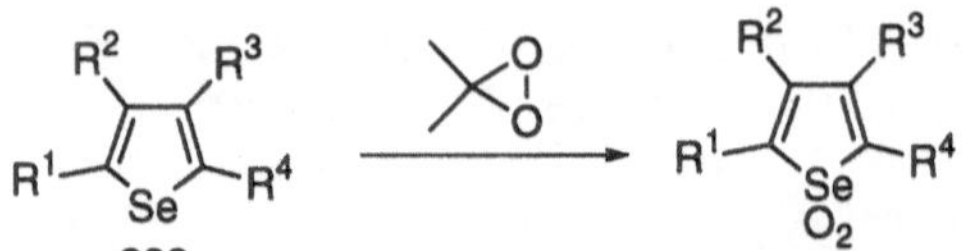

$R^1 = R^2 = R^3 = R^4 = Ph; 97\%, R^1 = R^2 = R^3 = R^4 = 4\text{-MeC}_6\text{H}_4; 89\%$
$R^1 = R^2 = R^3 = R^4 = 4\text{-MeOC}_6\text{H}_4; 99\%, R^1 = R^2 = R^3 = R^4 = 4\text{-ClC}_6\text{H}_4; 69\%$
$R^1 = R^4 = Me, R^2 = R^3 = Ph; 97\%, R^1 = R^3 = t\text{-Bu}, R^2 = R^4 = H; 97\%$

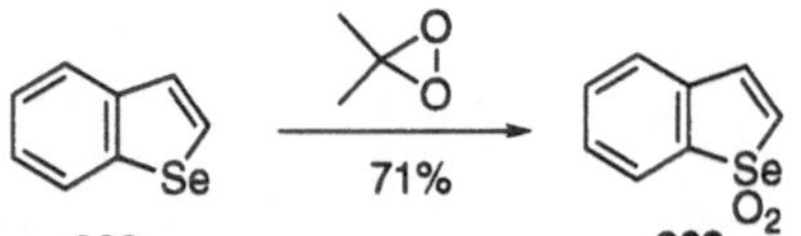

Scheme 108   **202**   71%   **203**

Selenophene 1,1-dioxides, **201** and **203**, are far less stable than the corresponding thiophene 1,1-dioxides, and decompose when heated at their melting points (135–155 °C). The dioxides **201** also decompose, when heated in refluxing toluene, to give the corresponding furans **193** in 50–80 % yields, probably with elimination of SeO (Scheme 109) [257]. ^{77}Se NMR signals of **201** and **203** appear in the region of δ 1018–1054, and in the IR spectra, symmetrical and unsymmetrical stretching vibrations of SeO$_2$ appear in the regions of 875–909 and 927–938 cm^{-1}, respectively [254, 255].

The chemistry of selenophene 1,1-dioxides is thus only now beginning to be investigated. Much yet remains to be discovered, in the development of a fruitful field of heterocyclic and heteroatom chemistry.

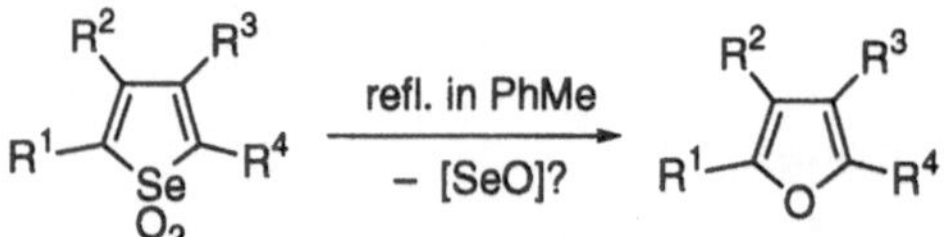

Scheme 109   **201**   **193**

5
Structural and Theoretical Study

Although the authors have collected all of the papers on the chemistry of thiophene 1,1-dioxides, the purpose here is to review its chemistry from synthetic aspects. Structural and theoretical chemistry of thiophene 1,1-dioxides is therefore not reviewed here. References are however given below for convenience.

X-*Ray single crystal structure analyses* of thiophene 1,1-dioxides (Fig. 2) have been reported. *Dipole moments* of thiophene 1,1-dioxides (Fig. 3) were experimentally determined or theoretically calculated.

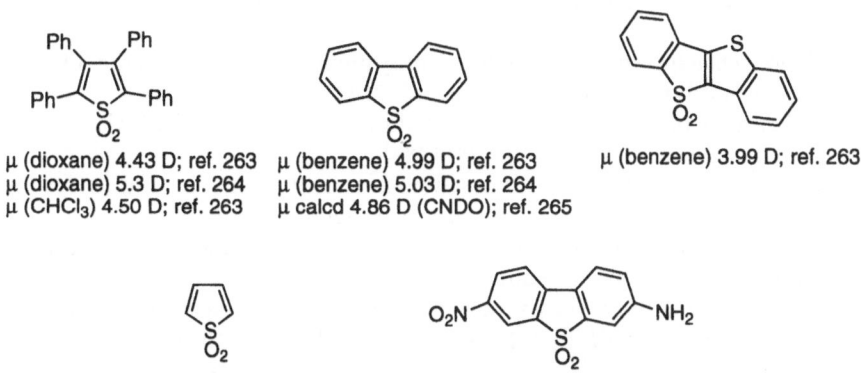

R^1 = R^2 = H; ref. 259
R^1 = Me, R^2 = H; ref. 260
R^1 = H, R^2 = Me; ref. 260
R^1 = Br, R^2 = H; ref. 259
R^1 = R^2 = Me; ref. 259

Fig. 2. Thiophene 1,1-dioxides for which X-ray single crystal structure analyses have been reported (with relevant references)

μ (dioxane) 4.43 D; ref. 263
μ (dioxane) 5.3 D; ref. 264
μ (CHCl$_3$) 4.50 D; ref. 263

μ (benzene) 4.99 D; ref. 263
μ (benzene) 5.03 D; ref. 264
μ calcd 4.86 D (CNDO); ref. 265

μ (benzene) 3.99 D; ref. 263

μ calcd (CNDO/2) 5.35 D; ref.266

μ calcd (CNDO) 9.55 D; ref. 265

Fig. 3. Thiophene 1,1-dioxides for which dipole mements have been experimentally determined or theoretically calculated (with relevant references)

A *polarographic* study has been made on a variety of monocyclic thiophene 1,1-dioxides, benzo[*b*]thiophene 1,1-dioxides, and dibenzothiophene 5,5-dioxides and their related compounds, and half-wave potentials were determined [267–271]. The examples given in Fig. 4 are representative [271].

Photoelectron spectra were determined to obtain vertical ionization potentials of 2,5-di-*tert*-butylthiophene 1,1-dioxide [272] and dibenzothiophene 5,5-dioxide [273].

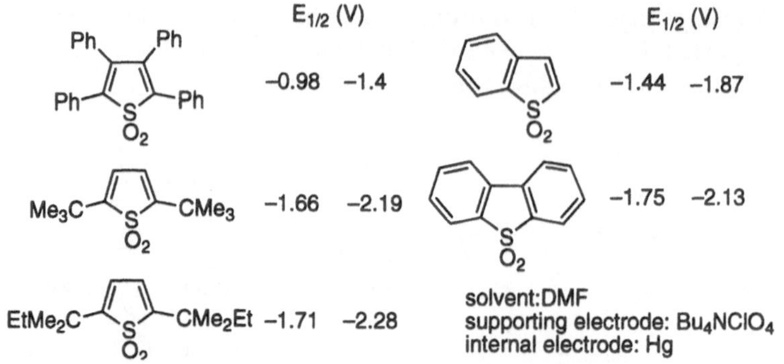

Fig. 4. Representative examples of monocyclic thiophene 1,1-dioxides, benzo[*b*]thiophene 1,1-dioxides, and dibenzothiophene 5,5-dioxides and their related compounds on which polarographic studies have been made and half-wave potentials determined (with relevant references)

ESCA study was carried out to determine ionization potentials of 2,5-dimethylthiophene 1,1-dioxide and the $Fe(CO)_3$ complexes of the parent thiophene 1,1-dioxide and 2,5-dimethylthiophene 1,1-dioxide [274].

The *UV/Vis spectrum* ($CHCl_3$) of the parent thiophene 1,1-dioxide shows two absorption maxima at 245 nm (ε 870) and 288 nm (ε 1070) [46, 58, 275]. UV/Vis spectra of tetraarylthiophene 1,1-dioxides [276], benzo[*b*]thiophene 1,1-dioxides [71, 277, 278], and dibenzothiophene 5,5-dioxides [278–280] have been reported.

Fluorescence spectra of 2-phenylbenzo[*b*]thiophene 1,1-dioxide and dibenzothiophene 5,5-dioxide have been reported [278].

ESR studies of the anion radicals of dibenzothiophene 5,5-dioxide and its derivatives have been carried out [281–289].

Mass spectra have been determined on a variety of thiophene dioxides and the fragmentation patterns are discussed in detail [290–301]. The mass spectrum of the parent thiophene 1,1-dioxide shows two strong peaks due to fragmentations to give thiophene and furan [46].

Raman spectra have been reported of only three compounds, 2,5-di-*tert*-butylthiophene, benzo[*b*]thiophene, and 3-methylbenzo[*b*]thiophene 1,1-dioxides [302].

IR spectral data on the SO_2 group of thiophene 1,1-dioxides are available in many papers. The $CHCl_3$ solution spectrum of the parent thiophene 1,1-dioxide shows two absorptions at 1306 cm^{-1} and 1152 cm^{-1} due to the SO_2 group [46]. In solution, unsymmetrical stretching vibrations of SO_2 group of thiophene dioxides appear in the range 1280–1332 cm^{-1}, and symmetrical ones in the range 1138–1170 cm^{-1} [53, 90, 303–306]. Solid state spectra (in many cases, for KBr disk and Nujol mull) show unsymmetrical stretching vibrations in the range 1265–1360 cm^{-1} and symmetrical ones in the range 1125–1190 cm^{-1} [22, 25, 37–40, 42, 43, 65, 68, 71, 85, 217, 302, 303, 307, 308]. The intermolecular hydrogen bonding ability of thiophene 1,1-dioxides to phenol and *p*-nitrophenol was investigated by IR spectroscopy [305, 309].

^1H *NMR spectral* data of a wide variety of thiophene 1,1-dioxides are available. Chemical shift values on ring protons of thiophene dioxides are given in [10, 33, 36–40, 42, 53, 54, 68, 90, 91, 217, 310–312]. The parent thiophene 1,1-dioxide shows two multiplets at δ 6.55–6.60 and 6.76–6.82 [46].

^{13}C *NMR spectral* data are also available in many papers [38–41, 312]. The parent thiophene 1,1-dioxide shows two signals at δ 128.7 and 130.9 [46]. Chemical shift values and assignment of each signal of benzo[*b*]thiophene 1,1-dioxides [313] and dibenzothiophene 5,5-dioxide [314] have been reported.

For *molecular orbital calculations*, see [265, 305, 315, 316].

6
References

1. For preceding reviews on thiophene 1,1-dioxides, see for example: (a) Rajappa S (1984) In: Bird CW, Cheeseman GWH (eds) Comprehensive heterocyclic chemistry, vol 4. Pergamon, Oxford, chap 3.14; (b) Rajappa S, Natekar NV (1997) In: Bird CW (ed) Comprehensive heterocyclic chemistry II, vol 2. Pergamon, Oxford, chap 2.10.3.2; (c) Raasch MS (1985) In: Gronowitz S (ed) Thiophene and its derivatives. Wiley, New York, p 571; (d) Simpkins NS (1993) Sulphones in organic synthesis. Pergamon, Oxford, p 319
2. For a review on thiophene 1-oxides, see Nakayama J, Sugihara Y (1997) Sulfur Rep 19:349
3. Backer HJ, Bolt CC, Stevens W (1937) Rec Trav Chim 56:1063
4. Backer HJ, Stevens W, van der Bij JR (1940) Rec Trav Chim 59:1141
5. Melles JL, Backer HJ (1953) Rec Trav Chim 72:314
6. Overberger CG, Mallon HJ, Fine R (1950) J Am Chem Soc 72:4958
7. Goldfarb YL, Kondakova MS (1952) Izv Akad Nauk SSSR Otedel Khim Nauk 1131
8. Schulte KE, Walker H, Rolf L (1967) Tetrahedron Lett 4819
9. Bordwell FG, Cutshall TW (1964) J Org Chem 29:2020
10. Neidlein R, Mrugowski E-P (1975) Arch Pharm 308:513
11. Cooper J, Scrowston RM (1971) J Chem Soc C 3052
12. Lamberton AH, McGrail PT (1963) J Chem Soc C 1776
13. Karaulova EN, Meilanova DS, Gal'pern GD (1958) Dokl Acad Nauk SSSR 123:99
14. Sunthankar AV, Tilak BD (1951) Proc Indian Acad Sci 33A:35
15. Rao DS, Tilak BD (1959) J Sci Industr Res (India) 18B:77
16. Sauter F, Sengstschmid G, Stütz P (1968) Monatsh Chem 99:1515
17. Bordwell FG, Albisetti CG Jr (1948) J Am Chem Soc 70:1955
18. Neumoyer CR, Amstutz ED (1947) J Am Chem Soc 69:1920
19. Gundermann KD, Fiedler G, Knöppel I (1976) Erdöl Kohle Erdgus Petrochem Brennst-Chem 29:23
20. Gilman H, Esmay DL (1954) J Am Chem Soc 76:5786
21. Davis W, Porter QN, Wilmshurst JR (1957) J Chem Soc 3366
22. Campaigne E, Hewitt L, Ashby J (1969) J Heterocycl Chem 6:553
23. Gilman H, Nobis JF (1945) J Am Chem Soc 67:1479
24. Block P Jr (1950) J Am Chem Soc 72:5641
25. Görlitzer K, Weber J (1980) Arch Pharm 314:76
26. Chapman NB, Hughes CG, Scrowston RM (1970) J Chem Soc C 2431
27. Foulger NJ, Wakefield BJ, MacBride JAH (1977) J Chem Research S 124
28. Clarke K, Fox WR, Scrowston RM (1980) J Chem Research S 33
29. Chambers RD, Cunningham JA, Spring DJ (1968) J Chem Soc C 1560
30. Chambers RD, Spring DJ (1971) Tetrahedron 27:669
31. Brooke GM, King R (1974) Tetrahedron 30:857
32. Venier CG, Squires TG, Chen Y-Y, Hussmann GP, Shei JC, Smith BF (1982) J Org Chem 47:3773

33. van Tilborg WJM (1976) Synth Commun 6:583
34. Nakayama J, Kuroda M, Hoshino M (1986) Heterocycles 24:1233
35. Raasch MS (1980) J Org Chem 45:856
36. McKeown NB, Chambrier I, Cook MJ (1990) J Chem Soc Perkin Trans 1 1169
37. Nakayama J, Yamaoka S, Hoshino M (1988) Tetrahedron Lett 29:1161
38. Nakayama J, Choi KS, Ishii A, Hoshino M (1990) Bull Chem Soc Jpn 63:1026
39. Nakayama J, Hasemi R (1990) J Am Chem Soc 112:5654
40. Nakayama J, Yoshimura K (1994) Tetrahedron Lett 35:2709
41. Krebs AW, Franken E, Müller M, Colberg H, Cholcha W, Wilken J, Ohrenberg J, Albrecht R, Weiss E (1992) Tetrahedron Lett 33:5947
42. Nicolaides DN (1976) Synthesis 675
43. Fujiwara AN, Acton EM, Goodman L (1969) J Heterocycl Chem 6:389
44. Balicki R, Kaczmarek L, Nantka-Namirski P (1993) J Prakt Chem 335:209
45. Miyahara Y, Inazu T (1990) Tetrahedron Lett 31:5955
46. Nakayama J, Nagasawa H, Sugihara Y, Ishii A (1997) J Am Chem Soc 119:9077
47. McKillop A, Kemp D (1989) Tetrahedron 45:3299
48. Rozen S, Bareket Y (1994) J Chem Soc Chem Commun 1959
49. Gilman H, Jacoby AL, Pacevitz HA (1938) J Org Chem 3:120
50. Gilman H, Avakian S (1946) J Am Chem Soc 68 1514
51. Lee DG, Srinivasan NS (1982) Sulfur Lett 1:1
52. Ledlie MA, Howell IV (1976) Tetrahedron Lett 785
53. Chen C-S, Kawasaki T, Sakamoto M (1985) Chem Pharm Bull 33:5071
54. Sharma KS, Parshad R, Singh V (1979) Indian J Chem 17B:342
55. Backer HJ, Melles JL (1951) Proc Koninkl Nederland Acad Wetenschap 54B:340
56. Chou T-S, Hung SC, Tso H-H (1987) J Org Chem 52:3394
57. Chou T-S, Chen M-H (1987) Heterocycles 26:2829
58. Bailey WJ, Cummins EW (1954) J Am Chem Soc 76:1932
59. Bates HA, Smilowitz L, Lin J (1985) J Org Chem 50:899
60. Chou T-S, Chen M-M (1988) J Chin Chem Soc 35:373
61. Backer HJ, Strating J (1937) Rec Trav Chim 56:1069
62. Backer HJ, Strating J (1934) Rec Trav Chim 53:525
63. Houge-Frydrych CSV, Motherwell WB, O'Shea DM (1987) J Chem Soc Chem Commun 1819
64. Bluestone H, Rimber R, Berkey R, Mandel Z (1951) J Org Chem 26:2151
65. Raasch MS (1980) J Org Chem 45:2151
66. Kadtrov AK (1979) Dokl Akad Nauk Tadzh SSR 22:667
67. Davies W, James FC, Middleton S, Porter QN (1955) J Chem Soc 1565
68. Takarabe K, Kunitake T (1980) Polymer J 12:245
69. Bordwell FG, Peterson ML (1959) J Am Chem Soc 81:2000
70. Terent'ef AP, Gracheva RA (1961) Zhur Obshchei Khim 31:217
71. Bergmann ED, Meyer AM (1965) J Org Chem 30:2840
72. Bassin JP, Cremlyn RJ, Lynch JM, Swinbourne FJ (1993) Phosphorus Sulfur Silicon 78:55
73. Meyer RF (1966) J Heterocycl Chem 3:174
74. Arya VP, Shenoy SJ (1973) Indian J Chem 11:628
75. Melloni G, Modena G (1971) Int J Sulfur Chem 1A:125
76. Melloni G, Modena G (1972) J Chem Soc Perkin Trans 1 1355
77. Novi M, Garbarino G, Dell'Erba C (1984) J Org Chem 49:1799
78. Novi M, Dell'Erba C, Garbarino G, Sancassan F (1982) J Org Chem 47:2292
79. Novi M, Garbarino G, Dell'Erba C, Petrillo G (1984) J Chem Soc Chem Commun 1205
80. Novi M, Dell'Erba C, Garbarino G, Scarponi G, Capodaglio G (1984) J Chem Soc Perkin Trans 2 951
81. Novi M, Garbarino G, Petrillo G, Dell'Erba C (1987) J Chem Soc Perkin Trans 2 623
82. Eastman RH, Wagner RM (1949) J Am Chem Soc 71:4089
83. Overberger CG, Hoyt JM (1951) J Am Chem Soc 73:3305
84. Hollitzer O, Seewald A, Steglich W (1976) Angew Chem 88:480

85. Ried W, Bellinger O, Oremek G (1980) Chem Ber 113:750
86. Cohen A, Smiles S (1930) J Chem Soc 406
87. Buggle K, Ghógaín UN, Nangle N, MacManus P (1983) J Chem Soc Perkin Trans 1 1427
88. Tominaga Y, Hidaki S, Matsuda Y, Kobayashi G, Sakemi K (1984) Yakugaku Zasshi 104:134
89. Harrison EA Jr, Rice KC, Rogers ME (1977) J Heterocycl Chem 14:909
90. Berestovitskaya VM, Titova MV, Perekalin VV (1977) Zh Org Khim 13:2454
91. Braverman S, Segev D (1974) J Am Chem Soc 96:1245
92. Wittig G, Ebel HF (1961) Liebigs Ann Chem 650:20
93. Pol VA, Kulkarini AB (1971) Indian J Chem 9:728
94. Pol VA, Kulkarini AB (1973) Indian J Chem 11:863
95. DeTar DF, Sagmanli SV (1950) J Chem Soc 965
96. Mustafa A, Zayed SMAD (1957) J Am Chem Soc 79:3500
97. Buggle K, O'Sullivan D (1974) Chem Ind (London) 343
98. Tamura Y, Ikeda H, Mukai C, Bayomi SMM, Ikeda M (1980) Chem Pharm Bull 28:3430
99. Braye EH, Hübel W, Caplier I (1961) J Am Chem Soc 83:4406
100. Gilman H, Swayampati DR (1957) J Am Chem Soc 79:208
101. Dent BR, Gainsford GJ (1989) Aust J Chem 42:1307
102. Bailey WJ, Cummins EW (1954) J Am Chem Soc 76:1936
103. Bluestone H, Bimber R, Berkley R, Mandel Z (1960) J Org Chem 26:346
104. Overberger CG, Whelan JM (1961) J Org Chem 26:4328
105. Gronowitz S, Nikitidis G, Hallberg A, Servin R (1988) J Org Chem 53:3351
106. Gronowitz S, Nikitidis G, Hallberg A (1991) Acta Chem Scad 45:632
107. Gronowitz S, Nikitidis G, Hallberg A, Stalhandske C (1991) Acta Chem Scand 45:636
108. Bordwell FG, McKellin WH, Babcock D (1951) J Am Chem Soc 73:5566
109. Davies W, Gamble NM, Savige WE (1952) J Chem Soc 4678
110. Davies W, Porter QN, Wilmshurst JR (1957) J Chem Soc 3366
111. Davies W, Ennis BC, Porter QN (1968) Aust J Chem 21:1571
112. Davies W, Porter QN (1957) J Chem Soc 826
113. Mustafa A (1955) Nature 175:992
114. Davies W, James FC (1955) J Chem Soc 314
115. Mustafa A, Zayed SMAD (1956) J Am Chem Soc 78:6174
116. Harpp DN, Heitner C (1970) J Org Chem 35:3256
117. Schloman WW Jr, Plummer BF (1976) J Am Chem Soc 98:3254
118. Davies W, Ennis BC, Mahavera C, Porter QN (1977) Aust J Chem 30:173
119. Amoudi MSEFE, Geneste P, Olivé JL (1981) New J Chem 5:251
120. Ikeda M, Uno T, Homma K, Ono K, Tamura Y (1980) Synth Commun 10:437
121. Eisch JJ, Galle JE, Hallenbeck LE (1982) J Org Chem 47:1608
122. Backer HJ, Dost N, Knotnerus J (1949) Rec Trav Chim 68:237
123. Albini FM, Ceva P, Mascherpa A, Albini E, Caramella P (1982) Tetrahedron 24:3629
124. Sauter F, Büyük G (1974) Monatsh Chem 105:550
125. Fischer U, Schneider F (1980) Helv Chim Acta 63:1719
126. Klärner F-G, Kleine AE, Oebels D, Scheidt F (1993) Tetrahedron Asymmetry 4:479
127. Sauter F, Büyük G (1974) Monatsh Chem 105:254
128. Bened A, Durand G, Pioch D, Geneste P, Declercq J-P, Germain G, Rambaud J, Roques R, Guimon C, Guillouzo GP (1982) J Org Chem 47:2461
129. Sauter F, Büyük G, Jordis U (1974) Monatsh Chem 105:869
130. Bougrin K, Soufiaoui M, Loupy A, Jacquault P (1995) New J Chem 19:213
131. Kabzinska K, Wróbel JT (1974) Bull Acad Pol Sci Ser Sci Chim 22:843
132. Bailey WJ, Cummins EW (1954) J Am Chem Soc 76:1940
133. Bluestone H, Bimber R, Berkley R, Mandel Z (1961) J Org Chem 26:346
134. Raasch MS (1980) J Org Chem 45:867
135. Davies W, Porter QN (1957) J Chem Soc 459
136. Dayagi S, Goldberg I, Shmueli U (1970) Tetrahedron 26:411
137. Christl M, Mattauch B (1985) Chem Ber 118:4203

138. Christl M, Freund S (1985) Chem Ber 118:979
139. Howard JAK, Mackenzie K, Johnson RE (1989) Tetrahedron Lett 30:5005
140. Mackenzie K, Howard JAK, Mason S, Gravett EC, Astin KB, Shi-Xiong L, Batsanov AS, Vlaovic D, Maher JP, Murray M, Kendrew D, Wilson C, Johnson RE, Preiß T, Gregory RJ (1993) J Chem Soc Perkin Trans 2 1211
141. Tobe Y, Kawaguchi M, Kakiuchi K, Naemura K (1993) J Am Chem Soc 115:1173
142. Kuck D, Bögge H (1986) J Am Chem Soc 108:8107
143. Kuck D (1994) Chem Ber 127:409
144. Ban T, Nagai K, Miyamoto Y, Harano K, Yasuda M, Kanematsu K (1982) J Org Chem 47:110
145. Hamon DPG, Spurr PR (1982) J Chem Soc Chem Commun 372
146. Spurr PR, Hamon DPG (1983) J Am Chem Soc 105:4734
147. Sedelmeier G, Fessner W-D, Pinkos R, Grund C, Murty BARC, Hunkler D, Rihs G, Fritz H, Krüger C, Prinzbach H (1986) Chem Ber 119:3442
148. Müller-Bötticher H, Fessner W-D, Melder J-P, Prinzbach H, Gries S, Irngartinger H (1993) Chem Ber 126:2275
149. Murty BARC, Pinkos R, Spurr PR, Fessner W-D, Lutz G, Fritz H, Hunkler D, Prinzbach H (1992) Chem Ber 125:1719
150. Wollenweber M, Hunkler D, Keller M, Knothe L, Prinzbach H (1993) Bull Soc Chim Fr 130:32
151. Thiergardt R, Keller M, Wollenweber M, Prinzbach H (1993) Tetrahedron Lett 34:3397
152. Craig DC, Paddon-Row MN, Patney HK (1986) Aust J Chem 39:1587
153. Beck K, Hünig S (1986) Angew Chem 98:193
154. Paddon-Row MN, Cotsaris E, Patney HK (1986) Tetrahedron 42:1779
155. Fessner W-D, Sedelmeier G, Knothe L, Prinzbach H, Rihs G, Yang Z-Z, Kovac B, Heilbronner E (1987) Helv Chim Acta 70:1816
156. Aitken RA, Cadogan JIG, Gosney I, Hamill BJ, McLaughlin LM (1982) J Chem Soc Chem Commun 1164
157. Akiba K-Y, Ohtani A, Yamamoto Y (1986) J Org Chem 51:5328
158. Jones DW (1973) J Chem Soc Perkin Trans 1 1951
159. Nakayama J, Yamaoka S, Nakanishi T, Hoshino M (1988) J Am Chem Soc 110:6598
160. Nakayama J, Hirashima A (1990) J Am Chem Soc 112:7648
161. Barton JW, Shepherd MK, Willis RJ (1986) J Chem Soc Perkin Trans 1 967
162. Fessner W-D, Sedelmeier G, Spurr PR, Rihs G, Prinzbach H (1987) J Am Chem Soc 109:4626
163. Mackenzie K, Proctor G, Woodnutt DJ (1987) Tetrahedron 43:5981
164. Melder J-P, Fritz H, Prinzbach H (1989) Angew Chem 101:309
165. Melder J-P, Prinzbach H (1991) Chem Ber 124:1271
166. Gupta YN, Houk KN (1985) Tetrahedron Lett 26:2607
167. Meth-Cohn O, Moore C (1987) S Afr Tydskr Chem 40:206
168. Dondoni A, Fogagnolo M, Mastellari A, Pedrini P (1986) Tetrahedron Lett 27:3915
169. Reinhoudt DN, Smael P, van Tilborg WJM, Visser JP (1973) Tetrahedron Lett 3755
170. van Tilborg WJM, Smael P, Visser JP, Kouwenhoven CG, Reinhoudt DN (1975) Rec Trav Chim Pays-Bas 94:85
171. Takeuchi K, Yokomichi Y, Kurosaki T, Kimura Y, Okamoto K (1979) Tetrahedron 35:949
172. Müller P, Schaller J-P (1989) Helv Chim Acta 72:1608
173. Barton JW, Lee DC, Shepherd MK (1985) J Chem Soc Perkin Trans 1 1407
174. Nakayama J, Hirashima A (1989) Heterocycles 29:1241
175. Meier H, Molz T, Merkle U, Echter T, Lorch M (1982) Liebigs Ann Chem 914
176. Scott LT, Cooney MJ, Otte C, Puls C, Haumann T, Boese R, Carroll PJ, Smith AB III, de Meijere A (1994) J Am Chem Soc 116:10275
177. Copland D, Leaver D, Menzies WB (1977) Tetrahedron Lett 639
178. Reiter SE, Dunn LC, Houk KN (1977) J Am Chem Soc 99:4199
179. Mukherjee D, Dunn LC, Houk KN (1979) J Am Chem Soc 101:251
180. Becker J, Wentrup C, Katz E, Zeller K-P (1980) J Am Chem Soc 102:5110

181. Zeller K-P, Berger S (1981) Z Naturforsch 36B:858
182. Wetzel A, Zeller K-P (1987) Z Naturforsch 42B:903
183. Neidlein R, Kohl M, Kramer W (1989) Helv Chim Acta 72:1311
184. Prochazka M, Horak V (1959) Coll Czech Chem Commun 24:2278
185. Gronowitz S, Nikitidis G, Hallberg A (1988) Chem Scr 28:289
186. Melles JL (1952) Rec Trav Chim 71:869
187. Wróbel JT, Kabzinska K (1974) Bull Acad Pol Sci Ser Sci Chim 22:129
188. Gronowitz S, Hallberg A, Nikitidis G (1987) Tetrahedron 43:4793
189. Bordwell FG, McKellin WH (1950) J Am Chem Soc 72:1985
190. Bordwell FG, Lampert BB, McKellin WH (1949) J Am Chem Soc 71:1702
191. Grandclaudon P, Lablache-Combier A (1983) J Org Chem 48 4129
192. Bordwell FG, Albisetti CJ Jr (1948) J Am Chem Soc 70:1558
193. Drozd VN, Sergeichuk VV (1977) Zh Org Khim 13:391
194. Sauter F, Jordis U (1974) Monatsh Chem 105:1252
195. Ried W, Mavunkal JB (1978) Chem Ber 111:1521
196. Udre VÉ, Lukevits ÉY (1973) Khim Geterotsikl Soedin 493
197. Udre VÉ, Lukevits ÉY (1977) Khim Geterotsikl Soedin 56
198. Udre VÉ, Voronkov MG (1972) Khim Geterotsikl Soedin 1602
199. Udre VÉ, Lukevits ÉY, Popelis YY (1975) Khim Geterotsikl Soedin 45
200. Udre VÉ, Lukevits ÉY, Kemme AA, Bleidelis YY (1980) Khim Geterotsikl Soedin 320
201. Buggle K, McManus P, O'Sullivan D (1978) J Chem Soc Perkin Trans 1 1136
202. Sauter F, Jordis U, Stanetty P, Hüttner G, Otruba L (1981) Arch Pharm 314:567
203. Ried W, Ochs W (1974) Liebigs Ann Chem 1248
204. Ried W, Mavunkal JB, Knorr H (1977) Chem Ber 110:1356
205. Ried W, Oremek G, Guryn R, Erle H-E (1980) Chem Ber 113:2818
206. Ried W, Oremek G, Guryn R (1981) Liebigs Ann Chem 612
207. Ried W, Pauli R (1984) Chem Ber 117:2779
208. Gronowitz S (1993) Phosphorus Sulfur Silicon 74:113
209. Karlsson JO, Gronowitz S, Hallberg A (1982) Chem Scr 20:37
210. Karlsson JO, Gronowitz S, Hallberg A (1982) Acta Chem Scand 36B:341
211. Svensson A, Karlsson JO, Hallberg A (1983) J Heterocycl Chem 20:729
212. Nikitidis G, Gronowitz S, Hallberg A, Stålhandske C (1991) J Org Chem 56:4064
213. Tsirk A, Gronowitz S, Hörnfeldt A-B (1995) Tetrahedron 51:7035
214. Squires TG, Venier CG, Hodgson BA, Chang LW, Davis FA, Panunto TW (1981) J Org Chem 46:2373
215. Aida T, Squires TG, Venier CG (1983) Tetrahedron Lett 24:3543
216. Gronowitz S, Bugarcic Z, Hörnfeldt A-B (1992) J Heterocycl Chem 29:1077
217. Kamiyama H, Nakayama J, unpublished results. See also: Kamiyama H, Hasemi R, Nakayama J (1993) Heteroatom Chem 4:445
218. Bordwell FG, Ross F, Weinstock J (1960) J Am Chem Soc 82:2878
219. Bordwell FG, Sokol PE, Spainhour JD (1960) J Am Chem Soc 82:2881
220. Bordwell FG, Hemwall RW, Schexnayder DA (1967) J Am Chem Soc 89:7144
221. Bordwell FG, Hemwall RW, Schexnayder DA (1968) J Org Chem 33:3226
222. Bordwell FG, Hemwall RW, Schexnayder DA (1968) J Org Chem 33:3233
223. Bordwell FG, Schexnayder DA (1968) J Org Chem 33:3236
224. Bordwell FG, Schexnayder DA (1968) J Org Chem 33:3240
225. Bordwell FG, Mecca TG (1972) J Am Chem Soc 94:5825
226. Schnorrenberg G, Steglich W (1979) Angew Chem 91:326
227. Schmidt H, Hollitzer O, Seewald A, Steglich W (1979) Chem Ber 112:727
228. Wild H, Mohrs K, Niewöhner U, Steglich W (1986) Liebigs Ann Chem 1548
229. Klein U, Mohrs K, Wild H, Steglich W (1987) Liebigs Ann Chem 485
230. Kirstgen R, Olbrich A, Rehwinkel H, Steglich W (1988) Liebigs Ann Chem 437
231. Geffken D, Groll G, Gleixner R (1987) Chem-Zeit 111:245
232. Hinsberg O (1915) Chem Ber 48:1611
233. Bordwell FG, McKellin WH (1951) J Am Chem Soc 73:2251

234. Brinon MC, de Bertorello MM, Bertorello HH (1983) An Soc Quin Argent 71:571
235. Becker S, Fort Y, Vanderesse R, Caubere P (1988) Tetrahedron Lett 29:2963
236. Chan M-C, Cheng K-M, Ho KM, Ng CT, Yam TM, Wang BSL, Luh T-Y (1988) J Org Chem 53:4466
237. Oae S, Makino S, Tsuchida Y (1973) Bull Chem Soc Jpn 46:650
238. Karaulova EN, Meilanova DS, Gal'pern GD (1960) Zhur Obshch Khim 30:3292
239. Robertson DW, Krushinski JH, Beedle EE, Wyss V, Pollock GD, Hayes JS (1986) Eur J Med Chem-Chim Ther 21:223
240. Ankner K, Lamm B, Simonet J (1977) Acta Chem Scad 31B:742
241. Shah KH, Tilak BD, Venkataraman K (1948) Proc Indian Acad Sci 28A:142
242. Suhr H, Wizemann S, Grünewald PHH, Iacocca D (1979) Acta Cient Venez 30:274
243. Grimshaw J, Trocha-Grimshaw J (1979) J Chem Soc Perkin Trans 1 799
244. Marmor S (1977) J Org Chem 42:2927
245. Nakayama J, Sugihara Y (1991) J Org Chem 56:4001
246. Nakayama J, Kamiyama H (1992) Tetrahedron Lett 33:7539
247. Takeshita H, Motomura H, Mametsuka H (1984) Bull Chem Soc Jpn 57:3156
248. McOmie JFW, Bullimore BK (1965) Chem Commun 63
249. van Tiborg WJM, Plomp R (1977) Rec Trav Chim Pays-Bas 96:282
250. Nakayama J, Sugihara Y, Terada K, Clennan EL (1996) Tetrahedron Lett 31:4473
251. van Tilborg WJM, Plomp R (1977) J Chem Soc Chem Commun 130
252. Fields EK, Meyerson S (1966) J Am Chem Soc 88:2836
253. Chambers RD, Cunningham JA (1967) Chem Commun 583
254. Nakayama J, Matsui T, Sugihara Y, Ishii A, Kumakura S (1996) Chem Lett 269
255. Matsui T, Nakayama J, Sato N, Sugihara Y, Ishii A, Kumakura S (1996) Phosphorus Sulfur Silicon 118:227
256. Nakayama J, Matsui T, Sato N (1955) Chem Lett 485
257. Umezawa T, Matsui T, Sugihara Y, Ishii A, Nakayama J (1998) Heterocycles 48:61
258. Douglas G, Frampton CS, Muir KW (1993) Acta Cryst C 49:1197
259. Faghi MSEAE, Geneste P, Olivé JL, Dubourg A, Rambaud J, Declercq J-P (1987) Acta Cryst C 43:2421
260. Faghi MSEAE, Geneste P, Olivé JL, Rambaud J, Declercq J-P (1988) Acta Cryst C 44:498
261. Goldberg I, Shmueli U (1971) Acta Cryst B 27:2173
262. Kronfeld LR, Sass RL (1968) Acta Cryst B 24:981
263. Gruntfest MG, Kolodyazhnyi YV, Udre VÉ, Voronkov MG, Osipov OA (1970) Khim Geterotsikl Soedin 6:448
264. Lumbroso H, Montaudo G (1964) Bull Soc Chim Fr 2119
265. Morley JO, Docherty VJ, Pugh D (1987) J Chem Soc Perkin Trans 2 1361
266. de Jong F, Noorduin AJ, Bouwman T, Janssen MJ (1974) Tetrahedron Lett 1209
267. Mazitova FN, Iglamova NA, Dmitrieva GV (1976) Khim Technol Topl Masel 13
268. de Jong F, Janssen M (1972) J Chem Soc Perkin Trans 2 572
269. Gerdil R (1973) Helv Chim Acta 56:196
270. Smith P, Sprague HG, Elmer OC (1953) Anal Chem 25:793
271. Iglamova NA, Mazitova FN, Vafina AA, Il'yasov AV (1979) Neftekhimiya 19:264
272. Müller C, Schweig A, Mock WL (1974) J Am Chem Soc 96:280
273. Solouki B, Bock H, Appel R (1975) Chem Ber 108:897
274. Eekhof JH, Hogeveen H, Kellogg RM, Sawatzky GA (1976) J Organomet Chem 111:349
275. Prochazka M (1965) Coll Czech Chem Commun 30:1158
276. Fortina L, Montaudo G (1960) Gazz Chim Ital 90:987
277. Zahradník R, Párkányi C, Horák V, Koutecky J (1963) Coll Czech Chem Commun 28:776
278. Dann O, Nickel P (1963) Ann 667:101
279. Mangini A, Passerini R (1954) Gazz Chim Ital 84:606
280. Morley JO, Docherty VJ, Pugh D (1987) J Chem Soc Perkin Trans 2 1361
281. Kaiser ET, Urberg MM, Eargle DH Jr (1966) J Am Chem Soc 88:1037
282. Gerdil R, Lucken EAC (1965) Mol Phys 9:529
283. Urberg MM, Tenpas C (1968) J Am Chem Soc 90:5477

284. Stasko A, Malík L, Tkác A, Pelikán P (1980) Z Phys Chem 261:1205
285. Stasko A, Malík L, Tkác A, Matasova E (1979) Zh Fiz Khim 53:486
286. Stasko A, Tkác A, Malík L, Adamcík V, Hronec M (1978) Chem Zvesti 32:294
287. Gerdil R, Lucken EAC (1965) J Am Chem Soc 87:213
288. Baiwir M (1971) Bull Soc Ry Sci Liege 40:162
289. Mispelter J, Grivet J-P, Baiwir M, Lhoste J-M (1972) Mol Phys 24:205
290. Bowie JH, Williams DH, Lawesson S-O, Madsen JØ, Nolde C, Schroll G (1966) Tetrahedron 22:3515
291. Blumenthal T, Bowie JH (1972) Org Mass Spectrom 6:1083
292. Bowie JH, White PY, Wilson JC, Larsson FCV, Lawesson S-O, Madsen JØ, Nolde C, Schroll G (1977) Org Mass Spectrom 12:191
293. Porter QN (1967) Aust J Chem 20:103
294. Fal'ko VS, Khvostenko VI, Udre VÉ, Voronkov MG (1971) Khim Geterotsikl Soedin 7:326
295. Dronov VI, Prokhorov GM, Syundyukova VK, Fal'ko VS (1971) Zh Org Khim 7:1751
296. Heiss J, Zeller K-P, Zeeh B (1968) Tetrahedron 24:3255
297. Budzikiewicz H, Gerhard U (1989) Org Mass Spectrom 24:699
298. Budzikiewicz H, Gerhard U (1992) Org Mass Spectrom 27:489
299. Vouros P (1975) J Heterocycl Chem 12:21
300. Bursey MM, Elwood TA, Rogerson PF (1969) Tetrehedron 25:605
301. Drabner G, Budzikiewicz H (1993) J Am Soc Mass Spectrom 4:949
302. Aleksanyan VT, Kumel'fel'd YM, Shostakovskii SM, L'vov AI (1965) Zh Prikl Spektroskopii Akad Nauk Belorussk SSR 3:355
303. Bavin PMG, Gray GW, Stephenson A (1960) Spectrochimica Acta 16:1312
304. Marziano N, Montaudo G (1961) Gazz Chim Ital 91:587
305. de Jong F, Janssen MJ (1973) Rec Trav Chim 92:1073
306. Tamura Y, Ikeda H, Mukai C, Bayomi SMM, Ikeda M (1980) Chem Pharm Bull 28:3430
307. Lamberton AH, Thorpe JE (1968) J Chem Soc C 2028
308. Pol VA, Kulkarni AB (1973) Indian J Chem 11:863
309. Kabzinska K, Wróbel JT (1974) Bull Acad Pol Sci Ser Sci Chim 22:181
310. Litvinov VP, Fraenkel G (1968) Izv Akad Nauk SSSR Ser Khim 1828
311. Numanov IU, Kadyrov AK, Komarov YI, Nasyrov IM (1976) Dokl Acad Nauk Tadzh SSR 19:34
312. Kabzinska K, Wróbel JT (1976) Bull Acad Pol Sci Ser Sci Chim 24:363
313. Geneste P, Olivé J-L, Ung SN, Faghi MSEAE (1979) J Org Chem 44 2887
314. Giraud J, Marzin C (1979) Org Magn Reson 12:647
315. Kanematsu K, Harano K, Dantsuji H (1981) Heterocycles 16:1145
316. Rozas I (1992) J Phys Org Chem 5:74

Author Index Volume 201–205